BACKYARD
FRUITS
·AND·
BERRIES

BACKYARD
FRUITS
◆ AND ◆
BERRIES

EVERYTHING YOU NEED TO KNOW
ABOUT PLANTING AND
GROWING FRUITS AND BERRIES
IN YOUR OWN BACKYARD

MIRANDA SMITH

RODALE PRESS, EMMAUS, PENNSYLVANIA

Quarto Publishing Staff:
Editor: Barbara Haynes
Senior Editor: Kate Kirby
Editorial Director: Sophie Collins
Book Designers: Neville Graham, Karin Skanberg
Picture Researcher: Susannah Jayes
Cover Designer: Bruce Low
Cover Photographer: Chas Wilder
Interior Illustrators: Ann Savage, Wayne Ford
Art Director: Moira Clinch

Rodale Press Staff:
Executive Editor: Margaret Lydic Balitas
Managing Editor: Barbara W. Ellis
Editor: Ellen Phillips

If you have any questions or comments concerning this book, please write to:
Rodale Press, Inc.
Book Readers' Service
33 East Minor Street
Emmaus, PA 18098

Library of Congress Cataloging-in-Publication Data
Smith, Miranda.
    Backyard fruits and berries : everything you need to know about planting and growing fruits and berries in your own backyard/Miranda Smith.
        p.   cm.
    Includes bibliographical references (p. 156) and index.
    ISBN 0–87596–638–1 hardcover
    1. Fruit-culture.   2. Berries.   I. Title.
SB355.S65   1994
634'.0484—dc20   93–50886
                                             CIP

Distributed in the book trade by St. Martin's Press
Typeset by Genesis Typesetting, Rochester, Kent, England
Manufactured by Bright Arts Pte. Ltd., Singapore
Printed by Star Standard Industries (Pte) Ltd, Singapore

2 4 6 8 10 9 7 5 3 1 hardcover

Contents

GETTING STARTED

Why do homegrown fruits taste so good? Freshness and ripening on the plant are two reasons, but growing techniques also play a part even if that part is mainly peace of mind. You can be certain your plants have the best possible soil and nutrition, making them full of vitamins, minerals and trace elements. You can also be certain that they have never been sprayed or treated with chemicals that could be harmful to yourself and your family. But the benefits of growing fruit go far beyond your health and palate.

Most fruits grow on beautiful plants. These trees and bushes add to your landscape's appeal in every season, from lovely spring-time blooms through summer's glossy foliage and fall fruit in all its shapes and colors. Even in winter, the plants' silhouettes can make a dramatic show.

OVERCOMING FEAR OF FRUIT GROWING

Growing fruit is as much a pleasure as eating it or looking at it. But despite this, gardeners are much more likely to grow vegetables or ornamentals. Three factors are responsible for gardeners' reluctance to grow fruit: time, space and fear of pests and diseases. Let's look at these factors more closely.

Time: Some fruiting plants, such as standard (full-size) apples or pears, take five to nine years to bear. The idea of caring for a food plant that long without a return puts

many people off. However, fruit cultivars grafted onto a dwarfing or semidwarfing rootstock will often yield harvests within two or three years. Other fruits, such as strawberries, kiwis and brambles, bear within a year or two of planting.

Space: A standard fruit tree can take up a great deal of space and cast an enormous shadow, but again, dwarfs and semidwarfs are much smaller. Other fruiting plants, such as strawberries, grapes and kiwi vines, conserve even more space. Having a good design and considering both the needs of the plants and the resources of your yard and garden, you can grow a surprising amount of fruit.

Pests and diseases: Horror stories about fruit crops with devastating insect pests and ugly diseases are as common, and often as exaggerated, as fish stories in Minnesota. And while it is true that huge single-crop orchards face serious pest and disease pressures, this is rarely true on a home garden scale, particularly if you garden organically.

When you have only a few fruiting plants, you can give them the time and attention they need to stay healthy and productive. And if you are already gardening organically, you have legions of beneficial insects and soil microorganisms that will come to your aid. But even if you're a beginning gardener or are growing on a previously uncultivated piece of land, you can avoid many problems with careful planning, planting and maintenance. This book will help you — and your fruits and berries — get off to a good start.

Chapter

1

CHOOSING AND PLANTING FRUITS AND BERRIES

No matter how small your yard, you can grow delicious fruits and berries. Breakthroughs in breeding have made fruit trees smaller and easier to harvest, so you don't need special equipment or a 50-foot ladder. And new styles of landscaping make it perfectly respectable to grow fruits and berries with your other ornamental plants rather than to hide them away in an orchard. In this chapter, you'll find out how easy it is to add fruits and berries to your own yard and learn how to decide which plants are best for you. You'll see how and where to site your plants by learning more about your yard's conditions. You'll find out how plants are sold, how to buy healthy plants and how to prepare an area for planting. And you'll learn how to plant fruits and berries to get them off to the best start.

Edible Landscaping

Gardeners are becoming more and more ingenious with their yard and garden designs. Edible landscaping—using food crops with other plants in the landscape planting rather than isolating them in a special area—was a new idea only a few years ago. Today, however, it is not uncommon to see blueberries tucked among azaleas flanking a front door, a small peach tree surrounded by flowers in a raised bed and grapevines curling up a trellis entryway.

If you are thinking about adding some fruit plants to your yard but have limited space, consider using them for both food and beauty. As long as you provide for the plant's needs, there is no end to the places you can find for fruits and berries. The list in "Fitting in Fruit" on page 13 is only the beginning—add to it according to your own situation.

MAKE FRUITS EASY TO GROW

A few simple guidelines can help you discover how easy and enjoyable it is to grow fruit plants. While you are planning your fruit garden, think about each of these measures.

Plan ahead: The majority of fruit crops are perennials. But unlike a clump of rudbeckia or a patch of veronica, fruit crops do not take well to being moved around the garden while you're finding the perfect spot for them. Since it is better to plant them only once, make good location choices before you buy the stock. "Permanent Places" on page 14 can help with this sort of planning.

Match plants to your climate: You wouldn't expect to see lemon trees dotting the green mountains of Vermont any more than you would imagine raspberries growing in a mangrove swamp. But you might think a 'Granny Smith' apple would fit into a New York garden or a 'Himrod' grape would grow in Georgia. Both choices would result in disappointment because they are climatic mismatches. To protect yourself against such mistakes, do some prebuying research. Find out what climate the species requires and, most important, the hardiness and other requirements of various cultivars. General information is included in the following pages as well as in the "Fruit Directory," beginning on page 98. Don't forget to check the USDA Plant Hardiness Zone Map on page 154 to find your hardiness zone.

▲ **Pruning becomes routine** *when you grow fruit at home. Here a gardener is removing water sprouts from a dormant apple tree. In the following pages, you will learn when and how to prune all the fruits you grow.*

▲ **Make a fruit tree a focal point** *in your yard. It will provide beauty through the year, shade in the summer and delicious fruit in season.*

Modify the plant's environment: Some fruiting crops need specialized environments for healthiest growth and highest yields. You can provide the correct conditions by planting them in an appropriate location or by giving them some extra care. For example, in New England, you might plant an early-flowering peach tree where it was sheltered from the cold, and in the warm, humid mid-Atlantic states, you would probably use a V-trellis to protect your raspberries from fungal diseases. Information about environmental conditions is given on page 16, as well as in the "Fruit Directory," beginning on page 98.

Prune and train well from the start: Pruning and training are imperative during the first few years of the lives of most fruiting plants. If you do this initial work well, the plants will be healthier and their subsequent care will be easier. Pruning and training information is given on pages 50–65.

Keep up with maintenance: Routine tasks such as watering and picking up dropped fruit are essential to success. Some of these chores, such as watering, decrease over the years, while others, such as picking up drops, aren't necessary before the plants bear. To avoid putting in more fruit plants than you can comfortably care for, plan for routine care from the beginning. You'll find maintenance chores discussed on pages 42–49 and in the "Fruit Directory," beginning on page 98.

Prepare for winter: In most sections of the country, the winter months bring dormancy — the plants cease to grow. If plants are not protected during this period, there is a danger of permanent damage. You can prevent most winter disasters with a minimum of time and energy. Winter protection of plants is discussed on page 49.

Observe your plants: Plant watching is an age-old and honorable occupation. A daily tour of your fruits and berries during the

▲ **Small trees,** *such as this dwarf pear, fit into mixed plantings in the backyard and produce about a bushel of fruit every year.*

▼ **Espaliered trees** *are the solution when you don't have much room. They cast little shade, yield well and add a touch of elegance to every landscape.*

▲ **Strawberries grow well** *in containers. They not only give high yields in very little space but also look good, especially when grouped together.*

▲ **Fig trees** *are attractive wherever they are grown. In the North, containerized figs should be moved inside for winter, while in the South, they can winter outdoors.*

growing season and a weekly tour during the cold months is the best way to get to know your plants. It is also good insurance. You'll know, for example, if Japanese beetles are getting out of hand on the brambles or if one of the plants has contracted crown gall or a virus. Observation can teach you when watering is necessary and whether or not to feed your plants. Plant watching is never a waste of time.

FITTING IN FRUIT

If you can't think of a place to put fruit on your property, consider these options:

Fruiting shrubs: Grow gooseberries, jostaberries or Juneberries instead of a fence along your property line. A living fence of brambles (raspberries and blackberries) will discourage neighboring dogs as effectively as a woven wire fence.

Strawberries: Grow strawberries in wooden barrels with holes drilled into the sides, in ceramic "strawberry jars" or in upright wooden pyramids.

Fig trees: Figs are ideal patio plants. They yield more prolifically when their roots are confined, so container growing suits them.

Grapevines and kiwis: Grapes will grow up almost anything, including the cast-iron supports for a porch roof, an arbor over your picnic table or the wires that form your fence. Kiwis are just as versatile.

Standard fruits: Grow full-size fuit trees in the front yard, where they make wonderful shade trees. Dwarfs and semidwarfs, some of which grow only 10 feet tall, can accent a perennial or shrub border or give a garden bench some shade.

Small fruit trees and spur-bearing shrubs: Prune and train dwarf fruit trees, currants and gooseberries to grow against a wall. Known as espaliers, they not only save space but also encourage early fruiting. You'll find more on this technique on pages 58–61.

Permanent Places

Most fruiting plants are perennials; you plant them once and they live in the same place for up to 50 years. With such long-term plants, it is vital to consider their ultimate size and needs. Just as vital is to think of your comfort and convenience. You will also want your plants to look as lovely as possible.

Some fruit plants, such as brambles, can be pruned to remain the same size from one season to the next, but most continue growing. Shrubs stabilize at a certain height but continue to put up new growth from their ever-widening root balls. To remain healthy and productive, shrubs require severe pruning at intervals during their lives. You must prune grapes so heavily to keep them disease-free and bearing well that you'll probably never see how large they are capable of growing. Trees reach their full height and width at different ages, depending on the species, cultivar and rootstock. After this, new growth is primarily limited to lateral branches and fruiting spurs.

PLANNING FOR SIZE AND SHAPE
Good planning is imperative when planting something that lives for many years. Before you put spade to soil, determine the exact location where your new plant will grow. Its ultimate size is one of the first things to consider. You will also want to estimate the mature plant's shade patterns so you can site the plant where its shade will be a blessing rather than a curse. For example, you might plant a tree growing on the west side of your house and garden, where it could shade a bed or two of cool-loving spinach and lettuce on hot afternoons, rather than on the east, where it might hide sun-loving vegetables from essential morning light.

Planning for mature size and shade patterns seems like an obvious thing to do. However, if you visit gardeners who have lived in the same place for a number of years, you will notice that it's easy to forget this caution. Many people overplant. It is common, for example, to see a group of three trees where there is only room for two. When you succumb to the temptation to overplant, you have only two alternatives: either live with a crowded

▲ **Standard pear trees** *can reach a height of 25 to 40 feet, depending on the cultivar and rootstock. Choose a site where their size and shape will be welcome for the many years of their life.*

situation and cope with the pest and disease problems that will probably result or grit your teeth and remove the extra plant. But unless you dig the plant out while it is young, root pruning for a couple of years first, the chances are good that it will die if you try to transplant it.

Fortunately, there are guidelines for spacing fruit plants. Good suppliers will give you this information when you purchase the stock, but if you're stuck, use the general rule of giving trees as much space between their trunks as they are going to be tall. For example, a 20-foot tree should grow no closer than 20 feet from its neighbors. Leave 15 feet between semidwarf cultivars promised to grow no taller than that. If you are planning to grow a standard beside a semidwarf, calculate spacing by the size of the standard. Spacing directions for other plants are given in the "Fruit Directory," beginning on page 98.

TREE TERMINOLOGY

When you look at a tree, try to identify its key parts. The drawing below shows features that you need to recognize when you are caring for your fruit trees.

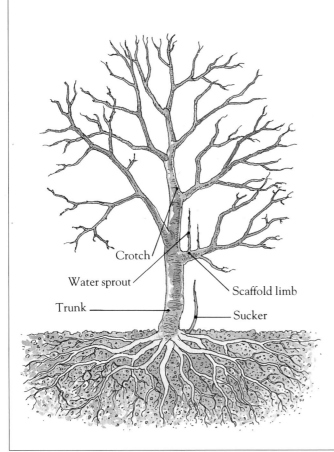

Crotch

Water sprout

Trunk

Scaffold limb

Sucker

Standard: A full-size fruit tree, usually maturing at about 20 feet in height.

Dwarf and semidwarf: Fruit trees grafted on size-controlling rootstocks. Dwarf trees often mature at 8 to 10 feet in height. Semidwarfs mature at 12 to 18 feet.

Genetic dwarf: A fruit tree that stays quite small without a dwarfing rootstock.

Rootstock: A cultivar onto which a fruiting cultivar is grafted. Rootstocks are selected for strong, healthy roots or for dwarfing effect.

Whip: A young tree, often the first-year growth from a graft or bud.

Scaffolds: The main structural branches .

Pome fruit: Fruit that has a core containing many seeds, such as apples and pears.

Stone fruit: Fruit with a single hard pit, such as cherries, plums and peaches.

Low-chill temperatures: Requiring fewer hours of cool temperatures to break dormancy.

High-chill temperatures: Requiring more hours of cool temperatures to break dormancy.

Self-fruitful: A tree that produces pollen that can pollinate its own flowers.

Compatible cultivars: Cultivars that can successfully cross-pollinate.

Crotch: The angle of emergence of a branch from the trunk.

Suckers: Shoots that sprout out of or near the base of a tree.

Water sprouts: Upright shoots that sprout from the trunk and main limbs of a tree.

From *Rodale's All-New Encyclopedia of Organic Gardening*
(Rodale Press, 1992)

PLANNING FOR BEAUTY AND CONVENIENCE

There's more to siting fruits and berries than size and shade. Since you're growing these plants in your own backyard, you'll want them to look good. If you create a landscape that pleases you, you're also likely to take better care of the plants in it. Consider beauty and convenience when you're trying to place a tree, shrub or vine. Start by trying to picture your yard as a whole rather than as a group of separate areas. Just as if you were positioning furniture in a room or flowers in an ornamental bed, you'll want to achieve a balanced look in your yard. Try to avoid weighing the landscape down by putting all the large elements, such as trees or large bushes, on one side of the yard. But if growing conditions force you to do this, you can add a counterweight with a bed of striking ornamentals on the opposite side. Whatever your taste, plan for it. The same tree can grow equally well surrounded by spring bulbs and low-growing flowers as it can standing alone in a bed covered with pea gravel.

Consider the down side: You might like the way a mulberry tree looks at the head of the driveway, but you won't appreciate splattered berries rotting on your car. You'll be unhappy stepping over June-dropped apples on the sidewalk or sweeping cherry pits off the patio. Remember to consider potential hazards, such as fallen fruit and petals or a concentration of bird droppings, while you plan. Again, you can generally avoid a lot of trouble by leaving 15 feet between a fruit tree and a sidewalk or patio area.

Convenience makes a difference: If you have to go to a lot of trouble to harvest or prune, you are likely to procrastinate. Initial design can have a big impact on later convenience. For example, if you've planted brambles or fruit-bearing shrubs to form a living fence, as mentioned previously, you'll need to prune and harvest from both sides. To allow for both plant growth and space to work, leave as much as 10 feet between the seedlings you plant and your property line. At this spacing, the plants can mature to full size without growing into your working space.

SOIL

Perennial plants should be planted in soil capable of sustaining them over the long term. Before planting, evaluate drainage characteristics, fertility, depth, pH and organic matter content. Few sites are blessed in all regards. However, you can minimize many problems by applying cultural or mechanical techniques.

Poor drainage: Soggy soils exclude the oxygen essential to both roots and micro-organisms. Cranberries are the only fruit plants that tolerate "wet feet." Soils that drain too quickly tend to be dry and nutrient-deficient. Native plums and bush cherries are the only fruits that do well in these poor conditions.

Poor drainage can be caused by a number of factors: too high a proportion of clay relative to sand and/or silt; compaction from heavy

---

ENVIRONMENTAL CONSIDERATIONS

Fruits require specific environmental conditions to remain healthy and bear well. When planning your fruit garden, take the following into account:

- Soil quality
- Drainage
- Speed and direction of the wind during each season
- Air drainage and circulation
- First and last frost dates
- Seasonal high and low temperatures
- Chilling hours
- Light availability and patterns
- Seasonal average relative humidity levels

SOLUTIONS TO DRAINAGE PROBLEMS

Raised beds *allow you to give roots space to grow above a hardpan or high water table. They can also make routine weeding easier.*

Containers *let you grow fruit on a deck. They can hold special soil mixes.*

Drainage ditches *can lead surface runoff away from plants that don't tolerate "wet feet."*

Green manure crops *are a source of organic matter and can break up hardpan layers.*

foot traffic or heavy machinery; a hardpan caused by frequent rotary tilling in wet conditions; a high water table; or buildup of runoff from a hill or building.

Wet clay soils can be improved over the long term by growing green manures and/or adding compost. If you don't want to wait the years required to improve your soil in this fashion, you can modify it when you plant, as described in "Putting Plants In" on page 32.

Don't plant fruits in areas where people tend to walk or drive. If you want to redesign your yard and need to put a fruit crop in an old pathway, build a raised bed. Raised beds can correct a multitude of problems. Use beds to increase drainage, improve soil and fight compaction. Choose raised beds if you're growing fruit in clay soils, over a hardpan or in the path of water runoff. Double digging creates the best bed possible (see page 29). However, it is a lot of work and not absolutely necessary unless you are contending with a hardpan. If the water table is high, a raised bed can sometimes help, but it might be wiser to forego planting deep-rooted fruit plants, such as trees, in favor of shallow-rooted brambles or shrubs, such as blueberries. For really waterlogged areas choose cranberries.

Farmers divert runoff with dry wells, drainage tiles and ditches. The Soil Conservation Service can tell you if either tiles or a dry well is necessary and, if a well is required, not only whether you can legally install one but also how to do it most efficiently.

If your problem is slight, you can put in a small drainage ditch without any complications. Dig the ditch where the running water will collect naturally. To prevent erosion, plant the banks with a soil-holding grass or crown vetch and line the bottom with rocks. Divert the water away from your yard and into a street sewer or natural catchment basin, but make sure that you aren't directing it toward your neighbor's land or over a sidewalk.

Excessive drainage: This problem can be as time-consuming to correct as bad drainage. Generally, coarse-textured, sandy or gravelly soils are to blame. Building up the humus levels in the soil with green manures and compost is the most effective solution. But if you want to plant right away, you can modify the soil in the planting area, taking care to improve it over a wide enough area so that roots will be encouraged to travel outward. Preparing sunken beds is also a solution.

Mulching will conserve moisture. The

effect of prevailing winds drying out the soil and increasing water loss from the leaves will be reduced if you can site the plants behind a windbreak, such as your house or garage.

Fertility: Soil fertility doesn't need to be as high for most fruits as it does for vegetables, which tend to be more shallow rooted and faster growing. Nonetheless, your plants will be more resistant to pests and diseases if you give them a steady supply of balanced nutrients. Specific recommendations are given in the "Fruit Directory," beginning on page 98, but, in general, stick to the organic gardener's method of giving soil organisms a good habitat and the materials they need to make nutrients available. Compost is the key to this strategy. Use it in your planting holes, and apply it every spring.

Soil depth: Soil depth is extremely important for all types of fruit crop. The roots of most plants are more extensive than the top growth. If your soil is only a thin covering over bedrock, roots will spread sideways rather than downward. This growth habit keeps the plants from anchoring themselves in the soil and makes them prone to toppling in high winds. Another problem with this growth habit is that it increases the water and nutrient demands the plants make on the surrounding area. If roots have only a shallow zone to grow in, they will compete with grass or other plants. Eventually, nutrient deficiencies or drought symptoms will cut yields and plant health. The trusty raised bed technique can come to your aid in this situation. However, remember that no raised bed is deep enough to sustain a standard tree or large semidwarf if there is little soil beneath it. Instead, plant small shrubs, brambles, vines or truly dwarf trees.

pH (soil acidity or alkalinity): The soil's pH measurement determines nutrient availability and uptake, which affect plant health and productivity. In general, fruit crops are no

▲ **Dwarf trees require** *slightly less soil depth than standards (full-size trees). Check with local nurseries to choose appropriate rootstocks for your cultivar, climate and soil type.*

different than vegetables or most ornamentals; they prefer a slightly acid soil, as close to a pH of 6.5 as possible. Blueberries and cranberries like a more acid soil — a pH of 4.5 suits them best.

Fortunately, you can change your soil's pH. A soil test, preferably from a university or private laboratory, can tell you your soil's pH and the quantities of limestone or sulfur required to adjust it.

As a rule of thumb, you can raise pH one point by adding 5 pounds of calcitic limestone or 7 pounds of dolomite per 100 square feet; lower pH one point by adding 1 pound of powdered elemental sulfur per 100 square feet. If your soil is a bit too alkaline or acidic, you can adjust the pH by simply adding organic matter.

Organic matter content: When soil organisms break down organic matter and release nutrients, the fine-textured material that remains is a soil conditioner known as humus. Humus holds water and nutrients in the root zone of the soil and improves drainage and aeration. Fallen leaves and dropped fruit form a layer of organic matter around plants in the wild. Since you will remove these materials to minimize pests and diseases in your garden, you will have to bring in organic matter to keep your soil healthy. Compost and organic mulches like straw and shredded leaves are good sources of organic matter. As a general rule, apply compost in the spring. You can cover it with an organic mulch or use a thick layer of compost as mulch depending on the amount of compost you have and how many plants you're growing. Remember not to use uncomposted manures, even when they are mixed with bedding, in the soil around fruit plants. As fresh manure decomposes, it can create nutrient imbalances or even burn the roots of your plants.

SUNSHINE

Full sun is critical for most fruiting plants to thrive and bear well. In general, save the sunniest spots in the yard for fruit. Your plants should get at least six hours of sunshine on a summer day. The only exceptions to this guideline are currants, gooseberries, lowbush blueberries and some bramble cultivars. If the only areas of the yard where you could grow fruits and berries are partly shaded, choose one of these fruits.

WIND AND AIR PATTERNS

Wind can help keep plants healthy by circulating air around them, decreasing humidity and reducing fungal infection. Unfortunately, it can also harm or even kill your plants. Spring winds can be quite cold, increasing the chance that blossoms will drop or freeze before they're pollinated. (Many apricot and peach cultivars will withstand fairly low winter temperatures, but because they bloom so early, the blossoms are vulnerable to cold winds.) During the summer, wind

▼ **Air drainage** *can make the difference between success and failure. Cold air sinks to the lowest level it can reach. Plants sited midway down a slope are slightly protected from spring and fall frosts, while* *those growing in low spots or frost pockets may suffer. Windbreaks, such as buildings and hedgerows, trap cold air on one side and give some frost protection on the other.*

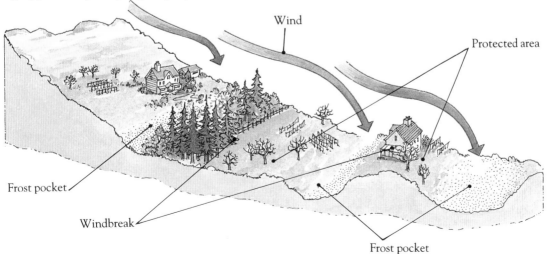

Wind

Protected area

Frost pocket

Windbreak

Frost pocket

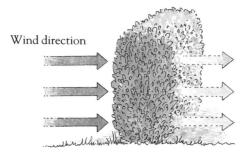

Wind direction

▲ **Hedges moderate wind** *by trapping some of the rushing air, thus slowing its passage. This is an advantage for plants that are marginally hardy, and it slows water loss from foliage. But it is a disadvantage for plants that are susceptible to diseases encouraged by high humidity.*

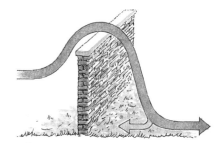

▲ **Walls and other solid barriers** *create a "windless" area on the leeward side. Use these spaces to protect early bloomers such as peaches, but remember that humidity is also apt to be high around plants growing against a solid wall.*

can bring cooling relief but may also dry out the plants, increasing watering needs. Sharp winter winds can kill plants that are unable to tolerate the resulting windchill.

Site your most tender plants in areas sheltered from the wind. Buildings, fences and hedges are all good windbreaks. You'll also find that bees do a better pollinating job in protected areas.

Air circulation and drainage: Many diseases and some pests thrive in humid environments, so you need to take air circulation and air drainage patterns into account when planning the location and spacing of your plants. In humid climates, experienced gardeners situate disease-prone fruit plants so that the prevailing wind blows across them. Commercial growers, who plant in rows, often set out plants in a north-to-south pattern or up and down a slope.

When you're deciding how far apart to space your plants, consider the average relative humidity levels in your area together with the strength of typical as well as unusual wind patterns. If you space plants as suggested by a nursery catalog or the "Fruit Directory," beginning on page 98, you will probably avoid excess humidity from overcrowding. However, surrounding buildings, trees or hills can make

▲ **Apricots, peaches and nectarines** *all bloom in the very early spring. If you live where late frosts can be a problem, plant these trees in areas sheltered from frosts and cold winds.*

an area essentially wind-free. If the climate is humid, this creates the same problem as overcrowding; you want air to move past the plants to dry off the foliage.

SPRING FROSTS

Late-spring frosts and freezing temperatures are a problem for fruits because they can kill blossoms, reducing your harvest. If your area is prone to late frosts, make sure you buy late-blossoming cultivars, especially if you grow peaches, plums or apricots.

SEASONAL HIGH AND LOW TEMPERATURES

High and low temperatures affect hardiness and ability to bear. In the South, gardeners worry that temperatures will be too warm for some species and cultivars to set fruit. Southern growers must also choose cultivars that are adapted to their long hot season.

Northern gardeners have to worry about winter lows. Many species and cultivars, including many apples, cherries and grapes, are adapted to very cold temperatures, but others, like oranges, will die. Mail-order catalogs include temperature tolerances when describing cultivars. However, as you'll find out in "Before You Buy" on page 22, temperature consistency is at least as important as temperature extremes.

WINTERKILL

Winterkill — when plants that seemed fine in fall don't survive the winter or have dead branches in spring — is actually a drought response. Although its effects are not visible until spring, it generally occurs in midwinter. Here's what happens: A warm spell with bright sunshine lasts long enough to thaw small limbs and branches. As soon as they have thawed, water begins to evaporate through the bark. But the plant's roots are in frozen soil; they can't take up moisture to replace the water being lost. The longer the warm spell, the more the plant dries out, so the more winter injury it sustains. Protect your plants during the winter as discussed in "Now They're Growing" on page 42.

HEATING AND CHILLING REQUIREMENTS

The number of hours above and below certain temperatures is critical to the performance of most fruit plants. Gardeners refer to these numbers as heating requirements and chilling requirements. Heating hours are calculated as the number of hours above 65°F that a plant must have to be healthy and bear well. For example, citrus and jujube trees have very long heating requirements and will not grow well in areas where they don't get their minimum hours.

Chilling requirements are more important to southern than northern gardeners. Chilling time is counted as the number of hours below 45°F that occur over the winter season. Chilling is important because plants are unable to break winter dormancy unless they have experienced a period of cool temperatures. If you plant a cultivar with a high chilling requirement in a southern area, the plant may be unable to break bud and start to grow in spring.

Wherever you live, contact your local Extension agent and find out your area's heating and chilling hours before buying plants. Then look up the heating and chilling requirements of the cultivars you want to buy, and choose the ones that will grow fruit reliably. Approximate chilling requirements, listed as "low," "moderate" and "high," are given for each entry in the "Fruit Directory," beginning on page 98. The ranges are as follows: low chill, 300–400 hours; moderate chill, 400–700 hours; high chill, 700–1,000 hours.

▲ **Ice gives some protection** *from frost damage. When late frosts threaten, growers often use the sprinkler to coat trees and plants near bud break.*

Before You Buy

An apple is an apple is an apple. . .or is it? It is from a botanical point of view, since every apple tree belongs to the same species. But cultivars of all fruits can be amazingly different from each other. Some prefer the cold snowy North, while others tolerate subtropical conditions. Some taste tart, others sweet. Some are crisp, others soft. Skin colors vary tremendously. And the differences go on and on.

CHOOSING PLANTS

Narrowing your plant selection—of species and then of cultivars—is one of the most challenging parts of starting a fruit garden. Unless you live in one of the coldest or hottest parts of the country—USDA Plant Hardiness Zones 3 or 10—you can grow most of the fruits mentioned in this book. Every year the selection widens as breeders develop cultivars that extend both the disease resistance and the hardiness range for various species.

Although it is theoretically possible to make good choices in an afternoon of impulse buying at a local nursery, it is far safer to do some planning and research before you buy. On page 26, you'll find a checklist of questions to ask before committing yourself to a particular plant or cultivar. Asking these questions can save time and trouble later on. Most are self-explanatory and others, such as the age of wood that bears fruit, will become clear as you read on. But you'll get the most from the checklist if you look over the following points first.

About the fruit: It is quite possible to buy a plant that does not bear (a sterile cultivar) or one that produces inedible fruit. Check before you purchase any fruit plant; some peaches, plums, cherries, crabapples and gooseberries are grown for decorative purposes only. It's also a waste of time, money and effort to grow a fruit you dislike. If at all possible, sample a cultivar's fruit before buying.

Climatic variation: Environmental conditions, particularly hardiness, are much more complicated than you would ever guess from a catalog. A cultivar bred and raised in one part of USDA Plant Hardiness Zone 3, for example, might not prosper in another section of the same zone. Parts of Zone 3, including sections of the upper Midwest and prairie provinces of Canada, experience consistently cold temperatures all winter. January and February thaws are infrequent and, if they do come, are short enough not to trigger spring growth. When spring does arrive, the warming trend is fairly steady.

Parts of the Northeast are also in Zone 3. In the Northeast, however, winter and early spring temperatures fluctuate and winter thaws are common. While temperatures may drop no lower than the tolerance listed in a cultivar's description, this variation can harm the plant by thawing tissues and stimulating the resumption of sap movement and budswell. Therefore, cultivar recommendations from neighbors and your local Extension agent are more reliable than numbers in a catalog.

It is also wise to remember that a zone map can't take your microclimate into account. If you live at a high elevation on a windy, north-facing slope in Zone 6, for example, it is possible that you are living in Zone 5

▲ **Match cultivars to climate**. *All these grapes can grow in the East or the Midwest. Here, clockwise from top left, are 'Beta', an early table (eating) and juice grape for Zones 5–8; 'Concord', a late table and juice grape for Zones 3–8; 'Seibel' cultivars, French-American hybrids grown in the East and West, for table and wine, in Zones 6–8; and 'Delaware', for table and wine in Zones 5–8.*

conditions, no matter what the map says. Similarly, if you live in a city located in Zone 5, your climate could be Zone 6 because the combination of heat-retaining concrete and windbreaks on every block can raise temperatures a full zone. So yes, use the map as a guide, but rely equally on your knowledge of conditions in your own backyard.

The self-fruitful and the infertile: Planting only one sweet cherry tree or a single kiwi vine can cause major disappointment. In the case of kiwis, most cultivars carry male and female blossoms on separate plants so you need two plants—a male and a female—to have fruit. In contrast, many flowers contain both male

THE NAME GAME

Botanical names include genus, species (sp.), sometimes subspecies (subsp.) or variety (var.) and cultivar (cv.), listed in that order. Members of a genus share certain botanical characteristics. For example, plums, peaches and nectarines, all members of the genus *Prunus*, share the same flower and fruit structures.

Species are related even more closely. All peaches are designated as *Prunus persica*, but in contrast, an edible plum can belong to one of at least three species: *Prunus insititis, P. domestica* or *P. salicina*. Nectarines are a natural mutation of a peach, and their botanical name, *P. persica* var. *nucipersica*, indicates this varietal status.

Cultivars are cultivated varieties—plants developed by breeders—and these names are generally in English and surrounded by single quotation marks, for example, 'Redhaven'. In principle, all 'Redhaven' peaches contain the same genetic material and are identical to each other. But that is not always the case, especially if a particular cultivar has been grown and propagated in the same location for a number of generations. Consequently, growers sometimes note the qualities of a particular "strain" of a cultivar.

Genus name — Species — Cultivar

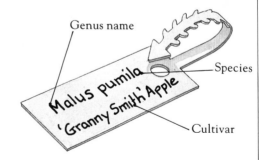

▲ **Plant tags list genus**, *species and cultivar names. Some tags include cultural information and environmental preferences.*

and female parts. They are called self-fruitful if they can fertilize themselves.

Some shrub and tree fruit cultivars are "self-infertile," meaning their pollen can't fertilize blossoms of the same cultivar. To get fruit, you must grow more than one cultivar of these plants. So when shopping, find out if the plant of your choice needs a pollinator and if so, choose one that blooms at the same time and also contains the same number of chromosomes. Good nurseries and mail-order catalogs will supply this information.

PARTS OF A FRUIT TREE

Less than a century ago, researchers at the East Malling Research Station in Kent, England, began to graft seedlings of standard (full-size) apple trees onto the roots of trees that were naturally small, or "genetic dwarfs." That experiment made backyard fruit growing a lot easier: The resulting tree was as small as the dwarf, but the fruit was identical to that of the full-size seedling.

Today, gardeners can choose from a range of dwarfing rootstocks for almost any tree fruit they choose to grow. The only limits are climatic: because dwarfing rootstocks are shallow rooted, they do not stand up well, literally, to the inevitable rigors of winter in Zones 3 and 4.

When grafted trees are planted, the graft union should be positioned several inches above the soil surface. If it does get buried, the scion is likely to send out its own roots, which will grow so vigorously that they will soon crowd out those of the more desirable rootstock. At that point, the tree will turn into a standard; the advantage of a dwarfing rootstock will be lost.

Interstem grafting: What happens if you want a dwarf and live in a cold-winter area? Rather than use a dwarfing rootstock that isn't hardy in a certain climate or under certain soil conditions, breeders will graft a stem—usually

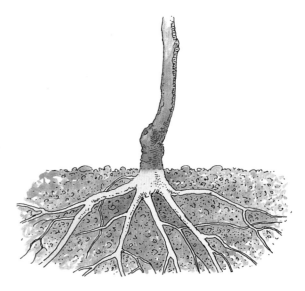

▲ **Graft unions** *are recognizable by the knobs they form on the trunk. Plant so that the union is at least 2 to 3 inches above the soil surface. If the union is buried, it may rot or the scion may form roots. Be careful not to cover the union with mulch unless directed to do so by your nursery.*

▲ **Plant containerized trees** *the same way you plant balled-and-burlapped and bareroot stock: their graft unions must be positioned well above the soil surface for the rootstock to confer its benefits— dwarfing or disease resistance—to the tree.*

of a dwarf or semidwarf tree—onto a hardy rootstock. The chosen cultivar is then grafted to this "interstem." Even though the rootstock may be from a standard tree, the dwarfing interstem gives its characteristics to the resulting tree. Interstems are normally buried to half their length when the tree is planted. However, there are exceptions to this rule. Remember to ask about planting depth when you buy the tree.

How Do They Come?

Your tree may come bareroot, balled-and-burlapped or grown in a container.

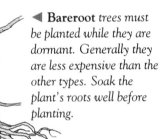

◀ **Bareroot** *trees must be planted while they are dormant. Generally they are less expensive than the other types. Soak the plant's roots well before planting.*

▶ **Balled-and-burlapped** *trees should also be planted when they are dormant. Check the root ball to be certain it is firm, and loosen the ties before burying the roots.*

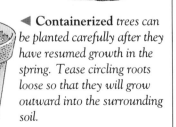

◀ **Containerized** *trees can be planted carefully after they have resumed growth in the spring. Tease circling roots loose so that they will grow outward into the surrounding soil.*

How does it come?

Most trees from mail-order suppliers arrive with their bare roots packed in moistened peat moss. Trees purchased from nurseries are more likely to come in containers or have balled-and-burlapped root balls. Each of these systems will produce a healthy tree if you handle the tree correctly when planting.

Sources of plants

You can buy plants either from a catalog or from a local nursery. There are advantages to either option and, of course, disadvantages, too.

Mail-order catalogs usually have a broader selection of cultivars than local nurseries. Some of the small catalog companies specialize in unusual plants, heirlooms or cultivars appropriate to a particular region. Many of these suppliers grow their own stock, and high-quality plants are the key to their success. You can rely on them to have taken the best possible care in both growing and transit. And as an added bonus, they are usually happy to answer any questions you might have and make suggestions about cultivars that are particularly appropriate to

▲ **Catalogs** *from mail-order suppliers list a broad selection of plants and cultivars, with information about hardiness zones and special qualities.*

ASK BEFORE YOU BUY

Take this checklist to the nursery with you, and go over these questions with the owner or a knowledgeable staff member before buying plants. You'll get the most from the checklist if you make a photocopy for each cultivar you're considering. Jot down the information to keep for reference.

ALL FRUIT PLANTS
Plant characteristics
- What size is the plant?
- What form (spreading, columnar or pyramidal) does the plant take with proper pruning and training?
- Does the cultivar bear on first-, second- or third-year growth?
- Will the cultivar perform well as a hedge? In a raised bed? In a lawn? On a trellis? As an espalier?

Fruit characteristics
- Does the cultivar bear edible fruit?
- Does the cultivar need pollinators of another gender or another cultivar to set adequate amounts of fruit?
- How long after planting is it before the cultivar bears fruit?
- Do I like the flavor, texture and appearance of the fruit?
- How large is the yield from this cultivar?
- Do I have to protect the fruit from birds?
- How involved and time-consuming is harvesting?
- Does the cultivar store, preserve and dry well?

Environmental information
- How typical are the environmental conditions of my yard to other places in this hardiness zone?
- Are the cultivar's chilling or heating requirements easy to meet in my environment?
- Is the cultivar hardy in my hardiness zone or one zone colder?

▲ **Stretch the grape season** *from late summer through the fall by growing several different cultivars that fruit at different times.*

- Is the cultivar grown in my neighborhood?
- Do neighboring gardeners who grow this cultivar commonly experience crop losses due to late or early frosts?

Maintenance information
- Do I need to prune and/or train in spring, in spring and summer or in fall?
- How complicated and/or time-consuming is the pruning or training, and does this change as the plant ages?
- How much water does the plant require during the years it is becoming established and when it is mature?
- Should the plant be mulched?
- What are the fertilizing recommendations for the plant?
- Do blossoms usually need frost protection?
- How do I winterize the plant?

Weeds, pests and diseases
- Are disease-resistant cultivars of the species available?
- What cultivars do local organic growers recommend?
- What pest and disease problems are common in this area with this species and cultivar?

▲ **Harvest times and flavors vary** *among raspberries. Plant at least two cultivars, and you can pick berries from July through October.*

- Are there growing techniques and botanical or biological controls for these pests and diseases that are both effective and easy to use?
- Is weed control important for this species and, if so, is preplanting a cover crop beneficial?

TREE FRUIT ONLY
Plant characteristics
- Does the cultivar fruit on spurs?
- Are genetic dwarfs or dwarfing or semidwarfing rootstocks available for the cultivar I want?
- What is the rootstock on this plant?
- What are the characteristics of this rootstock and is it hardy in my zone?
- Was an interstem used between rootstock and scion and, if so, what was it and what characteristics does it confer?
- If an interstem was used, should it be buried when I plant the tree and, if so, how deep should I bury it?

Fruit characteristics
- How long after planting is it before the cultivar bears fruit on this rootstock?
- How large is the yield from the cultivar/rootstock combination?

your area. The only potential disadvantage to buying from a small specialty company is that it may not have as broad a selection as the big mail-order houses.

Many large suppliers have earned their volume of business through a really good selection, superb plant quality and attention to customer relations. However, no matter how excellent the service and how good the plants, buying from a company outside your region is always a bit of a gamble. In general, it is a good idea to look for cultivars and stock in your own area first.

If you buy your plants from a local nursery, you may not have as many cultivars to choose from, but you will have several advantages. The plants at a reputable nursery will be grown in and adapted to local conditions, so you won't have to worry about hardiness. You can select each plant individually rather than settling for whatever you're sent. And you'll have a knowledgeable staff on hand to answer your questions and offer suggestions.

No matter where you buy your plants, check to see that the company guarantees them. These guarantees work on the honor system. They protect you from losing money on stock that dies in transit or has damage that was invisible until you start to plant. Needless to say, the guarantees do not cover plants whose trunks have been girdled by mice in your yard or have been run over by your lawn mower.

Help at hand: Help in choosing plants and cultivars is generally only a phone call away. People at area nurseries are usually both knowledgeable and informative. But if you suspect them of trying to drum up business, call your local Extension agent for advice. The "Resource Directory," beginning on page 155 lists organizations for fruit growers, most of which will be happy to send you information. Selecting plants may seem like a lot of effort at first, but before long, you will be enjoying the whole process.

Before You Plant

You've accounted for environmental factors, such as air, drainage and exposure. You've chosen a place where the plant will have space to grow and where it will look good. Now it is time to turn your attention to preparing the soil.

Advance preparation makes the job of planting easy. In most cases, it is sensible to prepare the planting area during the summer or fall before you put in the fruits. However, if you don't have the time, you can prepare most areas in the spring or fall you plant. As discussed on page 16, the soil should have good drainage, moderate fertility, high levels of organic matter and humus, and the correct pH.

DRAINAGE

Serious drainage problems should be corrected the year before you plant. This gives you time to fine-tune the corrections, if need be. Dig drainage ditches around the site you want to drain and watch the ditches through spring rains, summer storms and runoff to see how the water moves. Dig into the area to check how effectively the ditch is working. If you find that the soil is still soggy, you might want to increase the ditch size or change its location slightly.

In some cases, you may decide to build a raised bed to improve drainage. If you plan to plant your fruits in spring, you will need to double dig or build raised beds the previous year; otherwise, the soil may be too wet to work properly when planting time comes. In fall, covercrop these beds with oats or another annual that winter-kills. In the spring, you can clear away the dead stalks and dig your planting holes.

BUILDING SOIL FERTILITY AND HEALTH

Although you can't add much reserve fertility in a year, you can improve the availability of what is already in the soil. Because you are planting trees and shrubs, most of which like a deep soil with good organic matter content, plant a green manure crop such as hairy vetch or an annual clover in areas where you plan to grow fruits and berries. If you are going to plant in the spring, begin the summer before. However, if your climate allows fall planting, plant the green manure in the spring of the same year. You can apply compost or aged

---

COVER CROPS

Cover crops are used to protect soils from eroding, enhance nutrient availability, disrupt pest and pathogen habitats, and eliminate weeds. The annual cover crops listed below are good choices for areas where fruiting crops are to be the following year.

Crop	Seeding time	Seeding rate per 1,000 sq. ft.
Annual ryegrass	Spring, late summer	½–2 #
Buckwheat	Frost-free	2–3 #
Oats	Spring, late summer	3–4 #
Japanese millet	Frost-free	¾ #
Rapeseed	Midspring	¼ #
Terre verde alfalfa	Spring	½ #
Fava beans	Frost-free	2–4 #
Crimson clover	Spring	½–2 #

DOUBLE DIG A RAISED BED

Double digging a bed increases aeration and drainage. Follow these steps for good results. Begin several days before you plan to dig by marking off the bed and watering deeply. Use a sharp spade and start digging at one side.

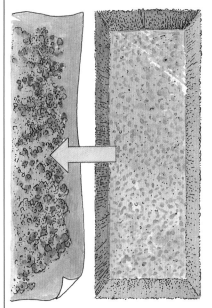

1 Make a trench *by removing about a cubic foot of topsoil across the width of the bed and placing it in a garden cart or on a tarp at the opposite side of the bed.*

2 Loosen the subsoil *after you have dug the trench. Push in a spading fork and wiggle it back and forth. Do this at close intervals all the way along the trench.*

3 Add well-finished compost *in a thin layer—between ½ and 1 inch—over the loosened subsoil if the topsoil is low in nutrients.*

4 Dig a second trench, *using the topsoil to cover the subsoil in the first trench. Follow steps 2 and 3 for each trench, covering the last trench's subsoil with that from the first.*

▶ **Raised beds will last longer** *if their corner boards are well secured. You can double dig the soil in them if drainage and nutrition need improvement. If not, you can simply add more topsoil or compost to the existing soil.*

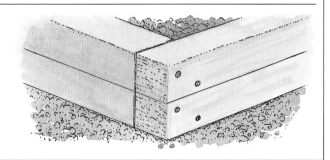

manure before you sow the green manure crop. As the green manure grows, it will take up fertility from the compost or manure and then, as it decomposes, return it to the soil.

If you wait until you prepare the area for planting before improving soil fertility, restrict yourself to compost or a compost-based fertilizer blend. But if the blend contains manure, make sure it's thoroughly composted—incompletely composted manure can burn roots.

Preplanting weed control: Weeds will always be with us. In the wild, they prevent soils from eroding and nourish and protect animals and microorganisms, including those that live in the soil and make it fertile. However, too high a weed population can seriously interfere with cultivated plants. Not only do weeds compete for water and nutrients but they can also harbor pests and diseases that prey as easily on your crops as they do on the weeds.

Weeds are not much of a problem to trees, particularly since a large area around the trunks should be kept clear and usually mulched with several inches of straw or wood chips. However, strawberries, brambles and small shrubs all perform better in areas where few weeds grow.

▲ **Quick-growing buckwheat** *is one of the most useful green manures because it shades out seedling weeds and grasses. It also makes phosphorus in the soil more available to crops.*

One standard definition of a weed is "a plant growing where you don't want it." This definition certainly applies to grass. Ordinary lawn grass can be a terrible weed. It is somewhat competitive but, more important, it is home to destructive grubs, like Japanese beetle grubs.

Hand-pulling and digging out grass roots is a real chore. Fortunately, there are easier ways to get rid of grass. The first job is to identify the growth pattern of the particular grass. If roots form a dense but compact layer of sod, like tall fescue or perennial ryegrass, you can simply cut under them and lift the sod away. A commercially available tool is ideal for this job. Electrically powered, it has two sharp, oscillating blades that cut no deeper than 8 inches into the soil. You rarely use them at that depth. Instead, you set them right under the sod layer. After several passes, you can lift up the sod as easily as lifting a layer off a cake.

Grasses that spread by rhizomes, like bluegrass or zoysiagrass, are more difficult. They can grow from the tiniest pieces of their extensive root systems. You pull on a root and pull some more until you finally break it, leaving a scrap to propagate itself in the soil. Completely removing these grasses by hand is almost impossible or enormously time-consuming, though stripping the sod off will remove most of the roots. If you are planning to plant brambles, bushes or strawberries where these grasses grow, let a natural system of cover-crop rotation do the job for you. This rotation also eliminates most of the annual weeds that plague home gardens.

Solarization: Soil solarization is an excellent technique for killing weeds, but it is effective only in sunny, warm climates. If you live in an area where summer skies are mostly bright and cloudless, you can solarize the soil during the summer before planting. Solarization kills many soil-dwelling pathogens, some soil-

Controlling Weeds

Green manures *smother weeds, add nutrients, feed microorganisms and improve soil structure.*

Electrically powered "hoes" *can lift a layer of grass roots or shallowly cultivate close to perennial plants.*

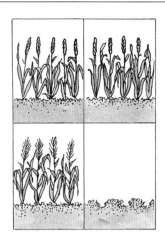

Rotate cover crops *Plant buckwheat after the last spring frost. Let it bloom, then mow closely. New plants will sprout. Till them in early fall. Plant winter rye and till in early spring. Plant fruit in two or three weeks.*

dwelling pests and some weed seeds in the top few inches of soil.

For tips and techniques to keep your plantings weed-free, see "Now They're Growing" on page 42.

Soil solarization depends on building high temperatures under a layer of clear plastic film. To solarize a bed, first mark off the area. Next, dig a 4- to 6-inch trench around the bed. Water the bed deeply and rake the surface smooth. Now cover the bed with a layer of 4- to 6-mil plastic film. Bury the film one side at a time so that you can pull it smooth and taut. Leave the plastic in place for a month or two during the very brightest, warmest months of the year. When you remove the plastic, the underlying soil will contain far fewer weed seeds and plant disease organisms.

Diseases in the wild: Many fruit diseases are carried from one plant to another by sucking insects such as aphids. The spores of others are blown by the wind, infecting any susceptible plant tissue they land on. Unless you live in the midst of a concrete, asphalt and glass landscape, fruit diseases are all around you.

Wild plants—abandoned apple trees, wild blackberries, native plums on the beach—carry all the diseases your plants could contract. Advisors generally suggest that gardeners cut down wildlings. However, real life is rarely that simple. You can't go around cutting trees on someone else's property any more than you can rearrange their cooking pots. Instead, cut down any wild plants on your own property, and take a census of where other fruit trees, bushes or brambles, both wild and domesticated, are growing. If they are on an abandoned lot close enough for bees to travel from them to your plantings, you might want to include them in your pest control strategy. If someone owns them, but allows pests and diseases to build to damaging levels, you might try to strike up cooperative workdays during appropriate seasons.

If you can't do anything about neighboring plants, take the best possible care of your plants and rest assured that a healthy plant growing in soil with good drainage and adequate nutrition is a resistant plant. And you'll find out how to prevent and cope with the pests and diseases that plague fruits and berries in "Coping with Pests and Diseases," beginning on page 72.

Putting Plants In

Planting is exciting. It is relative easy, too. Considering the amount of preplanting work you've done—from choosing locations and cultivars to preparing the site in advance—planting can seem like the simplest part of fruit growing.

Planting time is determined by climate. Early spring is the best time to plant in cool climates. You can work with stock that is reliably dormant (not in active growth), getting it into the soil in time for the roots to establish themselves before the buds swell and leaf out. Southern gardeners often plant in the fall. Because their soil does not freeze as rapidly or deeply as it does farther north, roots have time to grow over winter before topgrowth makes demands on them. If you are uncertain about the best timing for your region, ask area gardeners and your local Extension agent. It may be that you can plant certain species in the fall but should wait for an early spring date to plant others.

PREPLANTING CARE

Mail-order plants generally come bareroot. When they arrive, open the package immediately. Most suppliers transport the plants with plastic wrapped around the moistened peat surrounding the roots. Open the plastic to see how moist the peat moss is; it should neither be wringing wet nor bone dry. Most mail-order plants arrive in good shape, but this is the time to exercise your option on having plants replaced. If the peat moss was too moist when the plants were packed or the temperatures were too warm in transit, roots may have rotted. If they are slimy or covered with white or gray fungus, repack plants exactly as they came and send them back for replacement. Similarly, if the roots arrive dry, you will want to send the plants back. Check roots for dryness by bending a small rootlet; if it's

▶ **This carefully laid out** *and prepared planting area has paid off in the form of a peach orchard laden with blossom in springtime.*

▲ **Heeling-in** *saves roots from freezing or growing too soon by giving them the best possible storage conditions. Do not "plant" roots when heeling-in. Open the plastic wrap, but do not remove it or the peat. Pack soil around the roots to insulate them. Remove the soil carefully when planting time comes.*

totally dry, it will be brittle and snap off. If it is not brittle, a good soaking will probably revive the plant.

After checking the roots, your next steps depend on your planting schedule. If you are going to plant the following day, remove the plastic and set the roots in a bucket of lukewarm water. You need to soak them anywhere from 12 to 24 hours before planting. However, if you plan to wait for the weekend, sprinkle the peat moss with enough water so you can almost squeeze a drop out if you use all of your strength, then rewrap the roots in the plastic.

Complications: If you live in a snowy climate, your plants may come exactly on schedule, but the weather may not be conforming to this timetable. Obviously, you can't plant in frozen soil. But if the soil has been thawing and the weather is only suffering a temporary setback, you can store bareroot plants, still wrapped in plastic and moist peat moss, for about two weeks in a cool, dark area such as a garage. Don't let the roots freeze, and don't let them get warm enough to resume growth. Check the temper-

ature and the peat moss for moisture every day, adjusting as necessary.

Heeling-in areas: If plants arrive weeks before you think you can plant, you can "store" them or heel them in outdoors in a specially prepared area. Check the peat moss for moisture as described previously, and store the plants in a cool, dark place while you prepare the heeling-in area.

If there is snow on the ground, shovel it off of the planting area. Remove whatever unfrozen mulch you can and then lay clear plastic over the spot. Every day, remove thawed mulch until the plastic is lying on the soil surface. If you have already dug a trench and stockpiled soil, as suggested for cold-climate areas, you can now heel in the plants. If not, you need to wait until you can dig out enough soil to bury the roots.

Heel in plants by laying them at an angle, with their roots in the trench and trunks positioned at a slant so they neither lie along the soil surface nor stand upright. Heap soil over the roots and mulch with 4 to 6 inches of straw or hay. Encased like this, the plants are protected from freezing and will survive until you can plant them properly.

Balled-and-burlapped or container-grown plants: As you are unlikely to buy such plants until you are planning to plant, caring for them is easy. Simply keep them watered until you are ready to plant.

If, however, you do need to wait several days or a week before planting a container-grown or balled-and-burlapped plant, you will have to take special care since you cannot heel these plants in. If the plant is dormant, store it in a cool, dark area such as the basement or a garage to keep it from breaking dormancy and beginning to grow. Handle these plants carefully. Do not use the trunk as a handle. Instead, lift both balled-and-burlapped and container-grown plants under the root ball or by the pot. Even if the tree is firmly rooted, rough handling can set it back or injure it, since tiny roots are likely to be torn off or dislodged. Dormant, bareroot plants can take more rough handling than those with root balls. If you are planting a sizeable orchard, you might find bareroot trees easier to handle.

PLANTING FRUIT TREES

As illustrated on these two pages and page 36, planting fruit trees is relatively easy. However, how you plant can determine the health of the tree for its entire life, so you do need to pay attention to each step. Begin by assembling your tools: a sharp shovel, a hoe, a

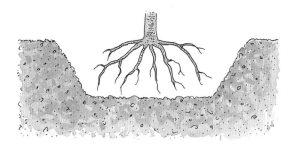

▲ **The perfect planting hole** *should be no deeper than the depth of the root ball and should measure twice the width of the roots across the top. Roughen up the sides, which should slope outward, so that the roots will be able to penetrate the surrounding soil more easily.*

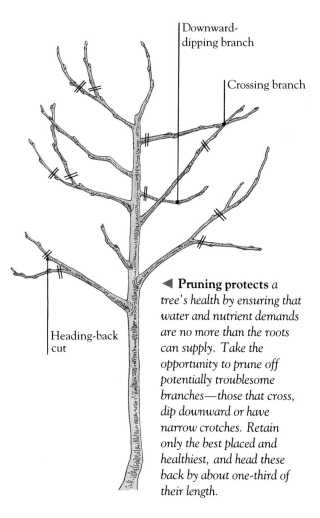

Downward-dipping branch

Crossing branch

Heading-back cut

◀ **Pruning protects** *a tree's health by ensuring that water and nutrient demands are no more than the roots can supply. Take the opportunity to prune off potentially troublesome branches—those that cross, dip downward or have narrow crotches. Retain only the best placed and healthiest, and head these back by about one-third of their length.*

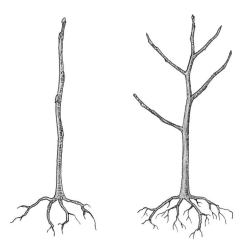

▲ **Prune at planting** *to determine the shape of the future tree. When you plant, look at your trees to make the best pruning decisions. Often, the trees you buy will have been pruned to an unbranched "whip" like the tree on the left. As new branches grow, you can choose the best. The tree on the right has a good framework and adequate roots to support the branches, so the branches will just need shortening.*

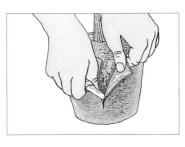

▲ **Remove containers** *before planting. Even peat pots can interfere with root growth. If the root ball doesn't slide out, cut the pot and peel it back.*

▲ **Root balls** *may be so dense that you have to use a knife to begin loosening the mass. Do this gently, cutting only as absolutely necessary.*

▲ **Circling roots** *must be loosened before planting. Try to tease them apart with your fingers, but if the mass is tightly packed, use a knife to begin the untangling.*

▶ **Details count . . .** *Pay attention when you plant. Left, the correct procedure; right, how not to go about it.*

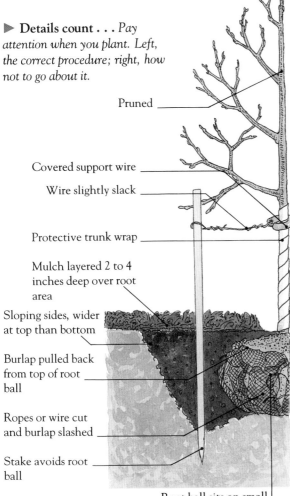

Pruned

Covered support wire

Wire slightly slack

Protective trunk wrap

Mulch layered 2 to 4 inches deep over root area

Sloping sides, wider at top than bottom

Burlap pulled back from top of root ball

Ropes or wire cut and burlap slashed

Stake avoids root ball

Root ball sits on small mound of firm soil

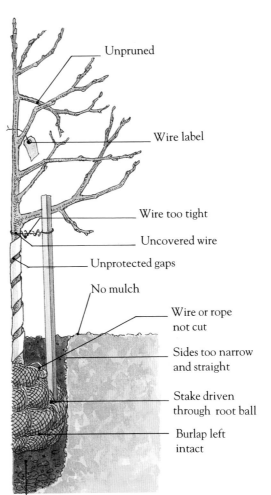

Unpruned

Wire label

Wire too tight

Uncovered wire

Unprotected gaps

No mulch

Wire or rope not cut

Sides too narrow and straight

Stake driven through root ball

Burlap left intact

Hole deeper than root ball; tree can sink even lower

pruner, a hose for watering, and a 5-gallon water bucket. For each tree you're planting, you'll need two tablespoons of liquid seaweed to put in the water bucket holding the roots, a tree guard, mulching materials and, if desired, stakes and cushioned wires to hold them around the trunk.

If you planted an annual crop the year before to eliminate weeds, you may already be working with bare ground. If the cover crop was a perennial such as winter rye, you'll need to till the area and let the crop residue rot for a week or two before planting. If you didn't use a cover crop and are planting into a lawn, you'll need to strip the sod from a 3 × 3-foot area for each tree.

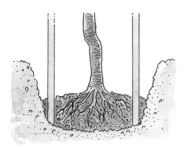

▲ **Bareroot trees** *are planted over a mound of soil in a hole with sloping sides. Take care to place the stakes so they don't pierce any roots.*

▲ **Containerized trees** *are easy to plant. After teasing the roots to separate them, set the root ball in the hole so the graft union is the correct height.*

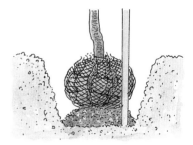

▲ **Balled-and-burlapped root balls** *should have their wrappings loosened before you bury them. Cut the wrappings away after setting the tree in place.*

▲ **Water the tree** *well when you have filled about three quarters of the planting hole. The water settles the soil, driving out air pockets.*

▲ **After the tree is planted,** *secure the trunk to stakes with cushioned wires. Leave a depression about a foot around the trunk to hold water, and lay mulch just beyond the depression.*

▲ **Fruit trees planted in lawns** *need mulch over their roots. As this tree grows, the gardener will continue to extend the mulched area to correspond to the drip line of the branches.*

PLANTING BUSHES

Planting bushes is no different than planting trees except that there is no need to worry about leaving a graft union above the soil or burying an interstem to half its length. Begin the operation by soaking bare roots in buckets of water. Some people use a liquid seaweed dilution—mixed at half the strength recommended on the bottle — instead of pure water. Roots should soak from 8 to 16 hours before you plant. Dig all your holes before you begin to plant. This gives you an opportunity to check to see that the placement and spacing are exactly what you want. Dig the holes just as you do for trees (see page 34). Now plant your bushes, burying them to the same depth they grew in the nursery. Don't forget to settle the soil with water as you backfill.

▲ **Careful planting** and thorough soil preparation help determine future plant health. This vigorous grapevine was planted correctly in fertile, well-drained soil a few years ago.

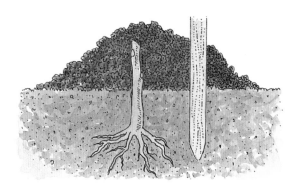

▲ **Newly rooted canes** are sometimes covered with loose mulch material or soil lightened with vermiculite when they are first planted. This allows the roots to become established before the buds break and start growing.

▲ **Good spacing** can help disease resistance by increasing air circulation. Use a zigzag pattern when planting double rows of bushes.

BRAMBLES

Brambles should always be planted in rows so they can be trained along a trellis.

VINES

Plant vines as if they were brambles — in rows for easy trellising.

STRAWBERRIES

Strawberries are usually planted in rows in the garden for easy maintenance. But if you only want a small crop, you can plant them in strawberry jars, wooden pyramids or tiers. In any case, remember that strawberries grow from the crown, where the stem and roots merge. If they are planted too deeply, first the new leaves and then the whole plant may die; planted too shallowly, they may wash out of the soil during a spring rain. All the roots must be buried, but the crown should remain above the soil surface.

PLANTING BRAMBLES

1 **Plant brambles deep enough** to bury new canes emerging from the crown. Dig holes or make a trench using a tiller attachment. Slope sides so hole is wider at the top than at the bottom.

2 **Add compost** to the bottom of the hole or trench. A layer of 1 to 2 inches is appropriate for most soils. Make sure you use finished compost or aged manure to avoid burning tender young roots.

3 **Place roots on a cone of soil.** Spread the roots gently with one hand while holding the stem in the other, and begin backfilling soil around the roots.

4 **Stop backfilling** when the roots are three-quarters covered. Then wiggle the stem gently to let some of the soil fall more closely around the roots.

5 **Water the planting area** to settle the soil even more. You want to eliminate any air pockets around the roots. Use a slow stream of water from a hose or bucket, watering all around the stem.

6 **Add more soil** to bring the nursery planting mark level with the surrounding soil. Water again to eliminate air pockets; add more soil if necessary. Finish by watering and leveling the area.

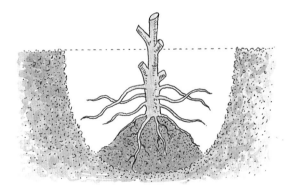

▲ **Plant grapevines** slightly deeper than they grew in the nursery in early spring. Place the roots on a mound of soil at the base of the hole. Add soil gradually, filling and watering well. Prune to a single stem with two buds.

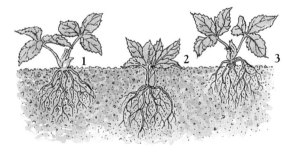

▲ **Strawberry crowns** must be above the soil line, but the roots must be buried: (1) correct depth; (2) too deep; (3) too shallow. Spread the roots when you plant. After planting, settle the plants with water to eliminate air pockets.

Chapter 2

GROWING AND HARVESTING YOUR FRUIT CROPS

❧

IN THIS CHAPTER, you'll learn the basics of watering, mulching, feeding, frost- and winter-protection techniques and good garden sanitation. You'll find out how to prune your trees, bushes, canes and vines to keep them vigorous and productive. You'll learn how to harvest, store and use your fruit crops. And, if you're inspired to grow more plants, you'll find the basic propagation techniques in an easy-to-follow, step-by-step format. In the wild, most plants die before maturing. Domesticated plants are equally vulnerable. The care you give them can make the difference between a long, healthy life and a shorter, less fruitful one.

Now They're Growing

*You've picked a site,
prepared the soil,
bought your plants and
put them in the ground.
What's the next step?
Those little whips,
canes, shrublets and crowns
will need some TLC from you to
become lush, productive plants.*

Fruit and berry plants need maintenance. Some require more than others, but all of them need adequate amounts of moisture and nutrients as well as protection from competitors and natural enemies.

WATERING

In most climates, gardeners have to water. "How much?" is the usual question, but simple answers, such as "1 inch a week" or its equivalent "2 gallons per square foot of root spread each week," are only guidelines. They help you remember to water and give you a ballpark amount. However, strictly adhering to such a formula assumes that all soil types have the same water-holding capacity, that sunshine, relative humidity and wind are a constant, and that all plants have equal needs. This is just not so.

Gardeners develop an eye for water stress. Wilting is the best-known and most obvious symptom, but long before plants wilt, their foliage loses gloss and the color loses vibrancy. This is the time to water, before the leaves droop.

Wilting is a sign of severe stress. It indicates that, for one reason or another, plant roots can't take up enough moisture to replace the

water that is being used and lost through evaporation. The plant protects itself by wilting—releasing a bit of water from each cell, including those that keep the stomates, or pores, open for easy movement of both water and air. Once this happens, the plant stops growing. It is unable to take in the atmospheric CO_2 necessary for photosynthesis, and it no longer has the cellular fluid necessary to metabolize sugars and manufacture other compounds. Wilting is a life-

▲ **Rain gauges** *give an accurate measurement of how much rain or water from a sprinkler has fallen. Keep regular rainfall and watering records to be certain that roots are receiving enough water.*

sustaining strategy; the plant is just trying to survive bad times. If the stress is not alleviated, the plant will eventually die. However, plants tend to die in fragments: first a leaf or two, then small branches and finally the entire plant succumbs to drought.

You aren't likely to let one of your plants die in this slow fashion. Nonetheless, even if you do use some kind of formula to make certain that you remember to water, teach yourself to watch foliage for signs of water stress as carefully as you watch the rain gauge.

Death by suffocation is a more common fate for new plants. If soils are kept too moist, without having a chance to drain and dry slightly between waterings, roots can suffer from lack of oxygen and moisture-loving pathogens can rot them.

WISE WATERING

- Keep track of natural rainfall with a rain gauge and water as needed.
- Apply water gently and slowly, using a soft spray from a water-breaker nozzle, a steady ooze from a soaker hose or with a drip irrigation system.
- Water deeply at weekly intervals. In extremely hot, dry conditions or with very sandy soils, you may need to increase frequency, but still water deeply every time.
- Water in the morning to diminish the rate of evaporation and to give foliage time to dry before dark, reducing the chance of fungal infection.
- For three years after planting, water as for first-year plants; blueberries and strawberries require this care all their lives.

WATERING OPTIONS

Watering time is a good opportunity to check on the overall health of your plants. Look carefully at new shoots as well as at older leaves and wood. Check the leaf color, look for insects and disease symptoms and make sure that the fruits are developing normally.

▶ **Watering by hand** *is appropriate only for small plants, since the moisture requirements of most fruit trees, bushes and vines are so great.*

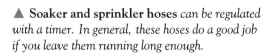

▲ **Soaker and sprinkler hoses** *can be regulated with a timer. In general, these hoses do a good job if you leave them running long enough.*

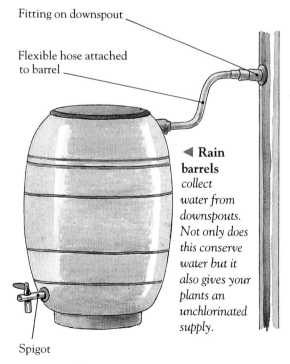

Fitting on downspout

Flexible hose attached to barrel

◀ **Rain barrels** *collect water from downspouts. Not only does this conserve water but it also gives your plants an unchlorinated supply.*

Spigot

MULCHING

Mulches are one of the gardener's best allies. They prevent weed growth, conserve soil moisture and insulate the soil from sudden temperature fluctuations. They also help prevent frost heaving and, depending on materials, add organic matter to the soil or increase soil temperature. Once you start using them, you discover that, in addition to all their practical attributes, mulches add to the physical beauty of your landscape. Mulches look tidy. They can also provide such a nice contrast of color and texture that some gardeners choose materials based as much on their appearance as on their beneficial qualities.

Mulching materials: Good mulching materials range from straw to pea gravel and black plastic. Each material has certain characteristics. Organic mulches such as straw, rice hulls and compost feed the beneficial soil micro-organisms that decay them. Like any other organic material, they improve soil structure and increase soil health.

Acid mulches such as pine needles, bark chips and shredded autumn leaves are most useful around blueberries, which need the added acidity. They also keep down weeds on pathways. Microorganisms that decompose these highly carbonaceous materials will use soil nitrogen to balance their nutrition. Consequently, if you use a high-carbon mulch over soils covering plant roots, guard against nitrogen deficiency by putting a 1-inch layer of finished compost underneath or sprinkling a high-nitrogen material like blood meal or cottonseed meal on the soil before adding mulch.

▼ **Strawberries fruit earlier** *when mulched with black plastic. However, fungal diseases can also thrive with this mulch material.*

Types of Mulch

Mulches can save you time and trouble. A well-chosen mulch can eliminate or minimize weeding, reduce irrigation needs, moderate soil temperatures and add organic matter to the soil.

Straw *reflects light back onto plants, absorbs water and is a good insulator. It is also easy to use and very attractive.*

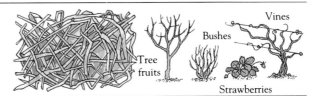

Gravel *discourages rodents, decreases soil water evaporation, minimizes weed growth and adds an "oriental" look to plantings.*

Dwarf fruit trees in containers and raised beds

Black plastic *eliminates most weeds from the planting area and raises soil temperatures a few degrees. It usually lasts for two years.*

Strawberries

Pine needles *increase soil acidity. Use them under blueberries or, if your soil is alkaline, under other fruit plants.*

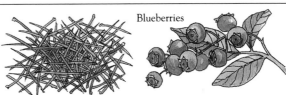

Bark chips *decompose slowly and give a finished look to a planting. But be warned: rodents love them! So leave 6 to 12 inches of bare soil around all trunks and stems.*

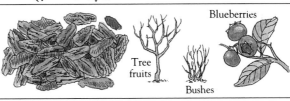

Shredded fall leaves *break down quickly into humus. Earthworms love them, too. But don't lay them on too thickly or they will exclude air— 6 inches should be enough.*

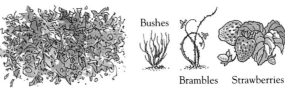

Compost *offers all the benefits of other mulches and also adds nutrients to the soil. A 3-inch layer suppresses most weeds.*

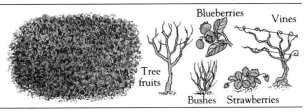

Inorganic mulches like plastics and gravel keep weeds down but do not add to the soil's health. Black plastic is particularly tricky to use well. Landscape cloth has gotten a lot of media attention recently as an ideal mulch. But growers have found that it is a poor choice for woody plants, since feeder roots grow up into the mulch.

Using mulches: Mulching depth is determined by the mulch material you use. Pile light, fibrous materials such as leaves, straw and spoiled hay 4 to 6 inches deep. Wood chips are more dense, so use no more than 3 inches. Rice hulls, compost, cocoa bean hulls and buckwheat hulls settle quickly, so a 3-inch layer is sufficient. Check the mulch depth over the course of the season and in the fall, replace the amount that has decomposed.

Mulches should not touch the trunk or stem of the plant. Besides encouraging rodents, some mulches can promote diseases because they hold moisture so well. Brambles, for example, are more prone to soilborne root and cane diseases when mulched. If you live in a humid climate, mulch them for winter protection, but remove the mulch in early spring. Wherever you live, keep a 6-inch circle of soil around tree, bush and vine trunks, mulching the soil outside the circle.

FERTILIZING

When you're dealing with fruits and fertilizers, it's easy to get too much of a good thing. Too much fertilizer, or nutrients applied at the wrong times, can hurt your plants. For safe, successful fertilizing, follow the guidelines outlined here.

If you suspect nutrient deficiencies in your soil, call your local Extension agent to ask about the location of the nearest laboratory that performs tissue tests. Commercial fruit growers use these tests because they give a more accurate reading of what the plant is actually taking up than a soil test, which

▲ **Straw** *is one of the best mulching materials. A few years under a straw mulch, and your soil will be rich and porous. Earthworms are partially responsible for this transformation.*

measures only the nutrients available to the plants. However, once the deficiency is corrected, test soil pH every two to three years, adding lime or sulfur as recommended.

Fertilize fruiting crops in the spring and early summer, and do not fertilize later than midsummer. Fertilizers may stimulate late growth that is too tender to survive winter conditions.

If you buy bagged fertilizers, choose well-balanced blends of organic materials such as rock phosphate, greensand, blood meal, feather meal and bonemeal. Like compost, these materials become available slowly and do not burn roots. Incorporate any fertilizers or soil amendments into the top few inches of

NUTRIENT CHOICES

Individual fruit plants vary in their nutrient requirements, but, fortunately, you don't need to adjust nutrient supplies precisely. Instead, test your soil to discover deficiencies and imbalances, and amend with slow-release organic materials such as the following.

Material	Nutrient	Analysis (NPK*)
Blood meal	Nitrogen	10–0–0
Bone meal	Phosphate	1–11–0; 24% calcium
Colloidal phosphate	Phosphate, potash	0–2–2
Feather meal	Nitrogen	11–0–0
Fish meal	Nitrogen	5–3–3
Kelp meal	Potash, trace elements	1.5–0.5–2.5
Rock phosphate	Phosphate, calcium, trace elements	0–3–0; 32% calcium, 11% trace elements
Sul-Po-Mag	Potash, magnesium	0–0–22; 11% magnesium, 22% sulfur

*Ratio of nitrogen to phosphate to potash

► **Backpack sprayers** *are convenient for spraying foliar feeds over a large area. Use several layers of cheesecloth to filter material going into the sprayer, since nozzles on these sprayers can easily clog. Spray just before sunrise or on a cloudy morning.*

the soil when you apply them and water immediately.

Compost is the mainstay of a good fertility program. It contains nutrients that are immediately available to your plants, but more than that, it provides insurance for the long term. When you apply compost, you are also adding microorganisms. Not only do they continue to break down the materials in the compost, but they also make nutrients already present in the soil available. Compost adds organic matter to the soil, improving drainage and water retention. It can also help balance soil pH, bringing both acidic and alkaline soils closer to neutral.

Foliar sprays: Many commercial fruit growers use foliar sprays to supplement nutrient supplies. Nutrients applied in minuscule water droplets move into the stomates (pores)

of the leaves and are immediately available for use. Typically, liquid seaweed, fish emulsion, fermented nettle tea and compost tea are sprayed on the leaves. Make fish emulsion and liquid seaweed sprays by diluting according to package directions. Fish emulsion is used solely as a nutrient source. Liquid seaweed adds trace elements and also confers some frost resistance. Fermented nettle tea adds large quantities of nitrogen and aids in disease resistance, while compost tea not only supplies balanced nutrients but also kills some plant pathogens and increases plants' resistance to others. For more information on these botanical sprays see "Preventing Diseases" on page 87.

FROST PROTECTION

Late spring frosts can kill tender blossoms and small fruit. In the fall, an early frost can damage or spoil ripening fruit, making it inedible or, if the frost was light, shortening its storage life.

Frost protection is always a gamble; sometimes it works, sometimes it doesn't, but it's worth a try. When frost seems likely in the spring, cover your fruit plants before evening dew coats their leaves and flowers. Old sheets, light blankets and stained tablecloths are all fairly effective coverings. Keep a good supply on hand. If you don't have a good supply of this sort of material, ask friends for castoffs and search yard sales and secondhand shops for them. Hay bales piled around the plants and covered with a sheet of plywood offer even more protection.

In the fall, be vigilant until all fruit is safely harvested. If frost threatens, harvest everything that will ripen as well indoors as out. Then cover plants to try to protect the remaining fruit.

SANITATION

Good sanitation is critical to success with fruit plants. Of all the elements in a pest and disease management program, sanitation is often the most effective measure. Codling moth larvae are a good example. When they burrow into an apple, it generally drops to the ground. If the drop is left there, the larva pupates, turns into an adult and repeats the cycle. However, if you pick up the dropped fruit and dispose of it in a compost pile hot enough to kill the larva or in a bag destined for the landfill, you have helped to diminish the population. This shows how important it is to pick up and dispose of all dropped fruit, including those that drop early in the season, especially if the fruit is infected with diseases or harboring pests. Also rake up fallen leaves from fruit plants and put them in a hot compost pile or otherwise dispose of them.

Here are more effective techniques to help keep pests and diseases at bay. Remove mulch from around diseased or pest-infected plants at the end of the season and compost it in a hot pile. Replace with fresh, clean mulch material. Renovate strawberry beds as suggested on page 152 in the "Fruit Directory." Prune off all diseased wood as soon as you see

GOOD SANITATION

Keeping your fruit plants clean becomes close to second nature once you discover how good sanitation practices lessen the incidence of pests and diseases.

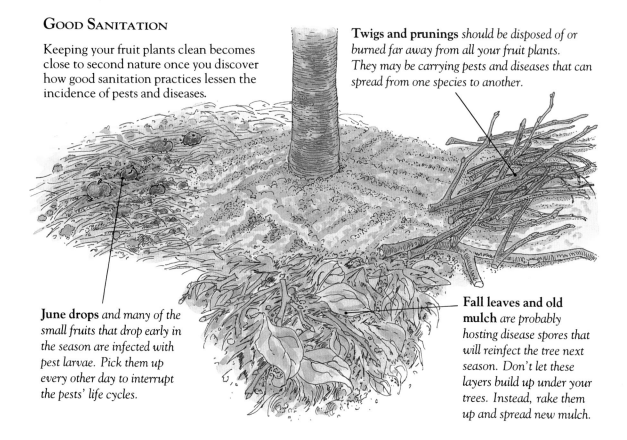

Twigs and prunings *should be disposed of or burned far away from all your fruit plants. They may be carrying pests and diseases that can spread from one species to another.*

June drops *and many of the small fruits that drop early in the season are infected with pest larvae. Pick them up every other day to interrupt the pests' life cycles.*

Fall leaves and old mulch *are probably hosting disease spores that will reinfect the tree next season. Don't let these layers build up under your trees. Instead, rake them up and spread new mulch.*

PROTECTIVE DEVICES

Protect your trees from weather and animals with some simple but essential devices.

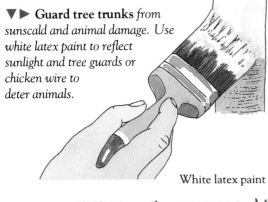

▼▶ **Guard tree trunks** *from sunscald and animal damage. Use white latex paint to reflect sunlight and tree guards or chicken wire to deter animals.*

White latex paint

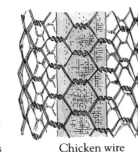

Tree guards   Chicken wire

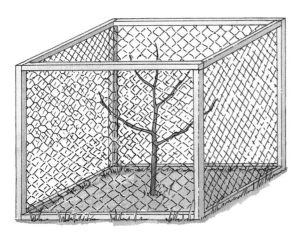

▲ **Cage young trees** *in deer country. A wire cage will keep deer from browsing on your trees no matter how deep the snow. Make cages in panels for easy disassembly and reuse elsewhere.*

it. Remove all prunings from the area and burn or bury them. Sterilize pruning tools between cuts with a 10 percent solution of laundry bleach.

WINTER PROTECTION

Just as the vegetable garden must be "put to bed" in the fall, perennial fruiting crops need some extra attention before winter. Use the following list for suggestions on winter-proofing your fruit plants, from both weather extremes and hungry mammals.

- Cover strawberry plants with mulch.
- Prune the canes that fruited on summer-bearing bramble plants and remove the prunings from the area.
- Mulch the soil around shrubs, vines and trees, taking care to leave 6 inches between trunks and stems and mulching materials.
- Wrap tree trunks with white plastic tree wrap or paint them with diluted white latex paint to protect against winter sunscald.
- Erect hardware cloth or aviary wire "fences" around tree trunks, about 6 inches from the trunk, to protect against winter rodent damage.
- Build a "deer cage" around young trees or shrubs. Use 2 × 4s and window screening for an inexpensive but effective barrier.

HOT COMPOST

Temperatures in a compost pile must be at least 145°F to kill pathogenic fungi and bacteria. Weed seeds are killed at 150°F, although some pathogenic viruses survive these temperatures.

Make sure your pile heats up by adding plenty of high-nitrogen materials like grass clippings, kitchen scraps and manure, keeping the pile evenly damp (it should be as moist as a squeezed-out sponge) and turning it every day or two to keep it aerated.

Pruning and Training

Pruning and training intimidates many beginning fruit growers. No wise words comfort them until they have actually taken pruners in hand and made that first cut. Fortunately, it won't be long before you discover that you don't have to be a rocket scientist, or even an experienced gardener, to prune and train well.

If you are terrified of making a pruning mistake, take heart. In the first place, pruning is not difficult as long as you pay attention to the pruning guidelines on these pages and use common sense both about the way the plant is already growing and the way you want it to grow. Second, remember that natural forces are constantly pruning — if not an ice storm, then a hurricane or maybe a deer. The point is that fruit and berry plants are organized to withstand pruning. Even if you do make a "mistake," it is more than likely that the plant will send out new growth to compensate.

Pruning provides several benefits to fruit plants. If well done, it strengthens the plant and makes it less vulnerable to damage from high winds and heavy snows. By reducing shading and competition, it encourages a higher production of good-quality fruit than unpruned trees would produce. It also removes diseased and insect-ridden growth, making the plant stronger and healthier. And it can prolong the life of a fruiting plant by allowing it to "renew" itself with young, healthy growth.

PRUNING AND PLANT HORMONES

To prune for maximum production of quality fruit, you need to know how plant hormones, or auxins, regulate plant growth patterns. Auxins made at branch and root tips confer "apical dominance," inhibiting the growth of other tissue such as buds lower on the limb. If you prune off the tip of the plant, dormant side buds will be free to grow. Each pruning cut you make will affect your plant's future growth on this invisible but potent hormonal level.

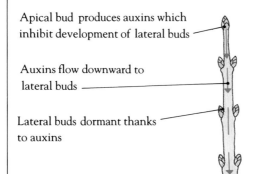

Apical bud produces auxins which inhibit development of lateral buds

Auxins flow downward to lateral buds

Lateral buds dormant thanks to auxins

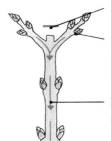

Apical bud removed

Lateral buds directly below site of removed apical bud lose dormancy

New dominant buds produce auxins that inhibit lower bud growth

PRUNING TOOLS

You will not need all of these tools at first. To begin with, choose a pruner, knife, loppers, a small saw and a long-handled saw. Then add other tools gradually.

▶ **Loppers and pruners** *come in a range of styles and sizes. Long-handled loppers are used to cut canes and branches less than 2 inches around. Anvil pruners have a sharp top blade and dull lower blade, whereas bypass pruners have two sharp blades, like scissors. Long-handled tree pruners extend your reach into high and awkward places on the plant.*

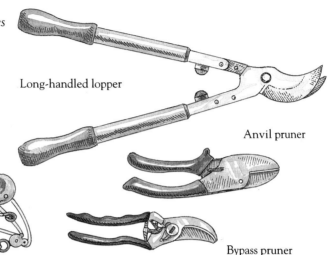

Long-handled lopper

Anvil pruner

Long-handled tree pruner

Bypass pruner

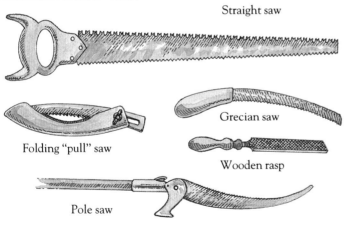

Straight saw

Folding "pull" saw

Grecian saw

Wooden rasp

Pole saw

◀ **For pruning thick branches** *you will need a straight saw. Folding "pull" saws cut on the pull stroke, while Grecian saws are curved to reach into small spaces. Pole saws reach into tall trees. A wooden rasp is used to smooth large cuts, which should then be painted with pruning compound to protect the wood.*

Pruning compound

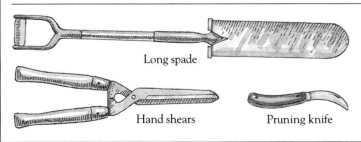

Long spade

Hand shears

Pruning knife

◀ **In addition**, *you will find a long spade useful for digging out suckers on brambles. Hand shears allow you to trim overly long growth. A pruning knife increases precision when making cuts and trimming away snags.*

KNOW WHAT TO PRUNE

Trees are individuals with their own particular characteristics and growth patterns. Good pruning means you have to work with the tree, bringing out its good points and pruning off its problem areas, rather than trying to make it conform to what might seem like the ideal fruit tree shape. After a few years of good pruning, you will begin to see the character of each of your trees. You will also benefit by having healthy trees that bear heavily year after year. To prune well, you must be able to identify what needs to be pruned and also know how to make good pruning cuts.

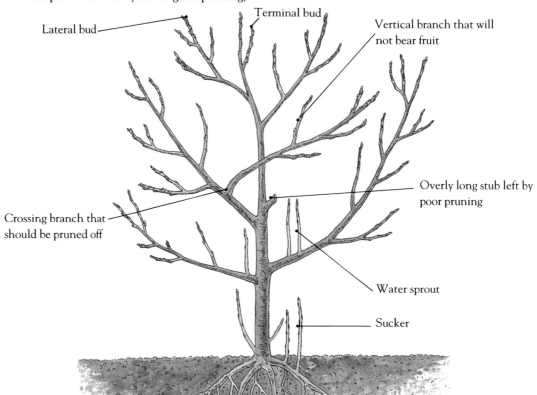

Lateral bud

Terminal bud

Vertical branch that will not bear fruit

Crossing branch that should be pruned off

Overly long stub left by poor pruning

Water sprout

Sucker

PRUNING TERMS

Branch collar: The part of the trunk that helps hold the branch to the trunk, often recognizable as a bulge at the base of the branch.

Branch crotch: The angle where a tree branch meets the trunk or parent stem.

Break bud: When a latent bud is stimulated into growing out into a leaf or twig, it is said to break bud.

Cane: A long, slender branch that usually originates directly from the roots.

Leader: The main, primary or tallest shoot of a tree trunk. Trees can be single-leadered, such as birch, or multiple-leadered, such as vine maple.

Espalier: A fruit tree trained to grow in a flat or virtually two-dimensional plane, usually against a wall or fence.

Pinching: Nipping out the end bud of a twig or stem with your fingertips to make the plant more compact and bushy.

Thinning cut: Cutting a limb off at the base either at ground level or at a branch collar.

Heading cut: Cutting a branch back to a side bud or shoot.

Sucker: An upright shoot growing from a root or graft union; also, in common usage, straight, rapid-growing shoots or water sprouts that grow in response to poor pruning.

From *Rodale's All-New Encyclopedia of Organic Gardening*
(Rodale Press, 1992)

PRUNING GUIDELINES

All pruning depends upon proper timing, tools and techniques. The following guidelines apply to all fruit plants.

- Learn about the growth habits of each fruit plant before you begin to prune.
- Choose the desired shape before you make the first cuts.
- *Think* before you cut; visualize the effect of each cut.
- Prune only as necessary to form the plant's framework and keep it healthy.
- Use your fingers to "rub out" unwanted developing buds during the season rather than letting them grow for the following year's pruning.
- Check the timing before you prune. For example, dormant pruning of apricots stimulates premature bloom, making them more vulnerable to damage from frost, but apples are strengthened by it. (See the "Fruit Directory," beginning on page 98, for specifics on each crop.)
- Choose tools appropriate to the job—use a saw to cut a branch, but use pruning shears to cut a twig.
- Sharpen tools before every use; sterilize them after every cut on a diseased plant.
- Cut all branches on a slant and generally just above an outward-facing bud.
- Do not prune frozen plants. If pruning in early spring, prune them after they have thawed but before they have resumed growth.

PRUNING AND TRAINING FRUIT TREES

In addition to the general pruning guidelines listed above, when you're pruning fruit trees the following points also apply:

- Cut all branches just beyond the "collar" (the supporting bulge at the base of a branch) to promote fast healing. Do not leave a stub and do not cut flush with the bark.

- When cutting back growth, make the cut above an outward-facing bud to promote spreading rather than crowding.
- When choosing limbs to form the framework of the tree, select those that form a wide (more than 40 degree) angle to the trunk. Remove those that form narrow angles (35 degrees or less), since they're prone to break off and will shade developing fruit.
- Choose scaffold, or framework, limbs that are about 6 to 8 inches apart on the young trunk and radiate around it.
- Remember that dormant pruning stimulates vegetative growth and summer pruning slows down growth and has a dwarfing effect.
- Make three cuts to remove all branches 2 inches or more thick.

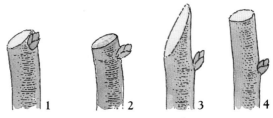

▲ **Proper cuts protect against diseases:** *(1) too close to the bud; (2) correct cut; (3) too far above the bud; (4) too long.*

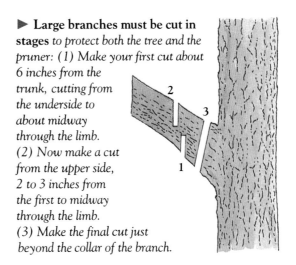

▶ **Large branches must be cut in stages** *to protect both the tree and the pruner: (1) Make your first cut about 6 inches from the trunk, cutting from the underside to about midway through the limb. (2) Now make a cut from the upper side, 2 to 3 inches from the first to midway through the limb. (3) Make the final cut just beyond the collar of the branch.*

Pruning at planting: Young trees must be pruned when they are planted. In general, container-grown and balled-and-burlapped trees need only have their topgrowth lightly pruned since their roots are intact. Bareroot stock, on the other hand, always loses a lot of root area when it is dug for delivery. This damage makes it impossible for roots to support as much growth as they did the previous year. To keep the plants healthy, topgrowth must be reduced. In some cases, nurseries prune before they send out stock. If they do so, they will caution you not to do any additional pruning when you plant. However, it is more common to receive a plant that needs to be pruned when it is planted. You can make one of two choices for pruning style at planting: either prune to a whip or leave some framework branches. The following year, prune again. This pruning job begins to establish the final shape of the tree. Tree shapes are discussed below and on page 58.

Third-season pruning repeats and emphasizes what you did in the second year. Prune lightly, since major pruning delays fruiting.

By the fourth year, the tree should definitely be shaped and scaffold branches not only well positioned but also strong and well formed. From here on, you'll only be pruning growth that obscures light from the center of the tree, prevents good air circulation, interferes with the pruning shape you have chosen or is infected with a disease, such as fire blight, that must be removed.

Training fruit trees: Pruning establishes the basic framework of the tree. However, some cultivars and some trees require training as well. Because the most fruits are borne on branches that are close to horizontal, it is necessary to train the branches to a horizontal position. If desirable scaffold branches grow upright, it can make more sense to spread them physically rather than remove them. Spacer bars, weights and even wooden clothes-

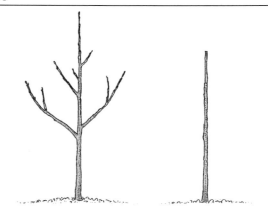

▲ **Prune to a whip** *if a bareroot tree does not have enough root area to support limbs.*

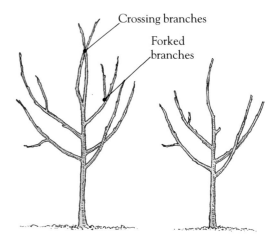

▲ **Prune off forks** *and crossing branches when you head back the branches; a fork left at the top of a tree can cause the trunk to split.*

▲ **A good scaffold structure** *is displayed by this tree. Careful pruning allows it to develop nicely.*

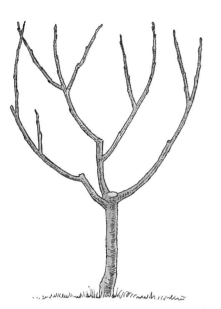

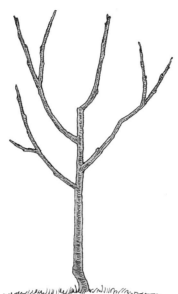

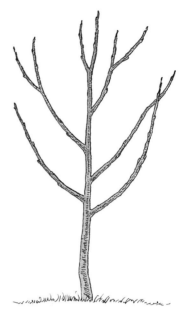

▲ **Vase, or open-center, pruning** *requires you to cut back the leader and any branches that threaten to become leaders. You want to encourage scaffold branches that radiate from an "open" center.*

▲ **Modified central leader pruning** *involves cutting out the central leader (central branch) once the fourth or fifth scaffold branch has formed. Topgrowth will then be easier to control.*

▲ **Central leader pruning** *allows the tree to gain some height. Keep scaffold branches that are well spaced and well positioned to let light into the center of the tree. Head back topgrowth if necessary.*

pins are all useful to encourage developing branches to spread.

If branches droop, you may have to prop them up. Fruit will not grow well on branch tips that dip toward the ground. These tips can be cut back to horizontal growth. However, if they are quite long and potentially productive, you may prefer to prop them rather than cut them back.

Thinning fruit: Most trees set more fruit than they can support. Pests such as plum curculio inadvertently cause some of the extras to drop, but it is still likely that the fruit set will still be too great to bear. If all fruit is allowed to grow and ripen, it is likely that the tree will respond by bearing lightly the following year, if at all. The weight of all that ripe fruit can break branches as well. In addition, pests such as codling moth are attracted to places where fruits touch, making a nice, snug niche.

Prune off, or thin, extra fruit just after it has set unless curculio pressure is extremely high. In that case, you can wait to thin until all infested fruits have dropped to the ground. Apples, pears and peaches should be thinned to a distance of about 6 to 8 inches between fruits or, if fruits are growing on spurs, to only one fruit per spur. Thin plums to one fruit every 3 to 4 inches. Cherries and quinces do not require thinning.

Pruning mature trees: Mature trees require only routine pruning if you have taken care of them over the years. Prune standard trees to keep topgrowth at a reasonable height, and cut back laterals when they get too large. With

PRUNING AN OLDER TREE

The roots of an old tree are accustomed to supporting a great deal of topgrowth. If too much of this topgrowth is removed at one time, the tree will respond by putting out an enormous number of watersprouts and suckers, as well as lots of new growth the following spring. If, on the other hand, you prune the tree over three to five years, following the guidelines given below, you won't have to deal with such an abundance of new material.

Pruning is not the only renovation chore; you need to rebuild the tree's health as well. Begin by spreading a circle of compost, starting about 18 inches from the trunk and extending out to the dripline (the edge of the tree's canopy). Water well all summer to make certain that the nutrients in the compost are leaching down to the roots. Thin extra fruit, control pests and diseases and follow normal maintenance and sanitation practices.

▶ **Year one:** *Knock off loose bark. Prune off broken branches and dead or diseased growth, cutting back to live wood in each case. Camouflage large wounds with pruning compound.*

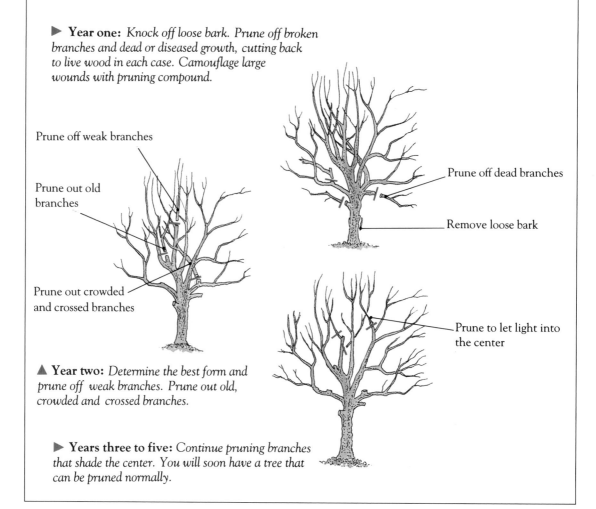

Prune off weak branches

Prune out old branches

Prune out crowded and crossed branches

Prune off dead branches

Remove loose bark

Prune to let light into the center

▲ **Year two:** *Determine the best form and prune off weak branches. Prune out old, crowded and crossed branches.*

▶ **Years three to five:** *Continue pruning branches that shade the center. You will soon have a tree that can be pruned normally.*

THE SPUR SYSTEM

Spurs form more readily if you let a lateral grow for a year, then prune back to the top flower bud in winter. Identify flower buds by their plump appearance.

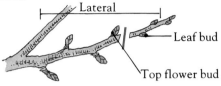

Cut back first-year branches *during dormant pruning, leaving four buds on each.*

Fruit grows *from these buds during the season. When you harvest, a short spur will remain.*

Young spur systems form *just below the fruit, generally increasing in number.*

As spur systems mature, *they may become too crowded. Prune off some of the older spurs.*

dwarfs and semidwarfs, pruning is done more often to promote tree health than to control size. Pruning mature trees becomes difficult only if they have been neglected. You may have moved into a house and discovered an old apple tree in the backyard, for example. Some neglected trees are too old and diseased to try to save, but most are worth renovating.

Spur pruning: Many cultivars of apple, pear, plum and cherry bear on short fruiting twigs called spurs rather than on the limbs. In addition to pruning extra fruit from each spur, you can also prune off some of the spurs. Removing extra spurs encourages the remaining ones to produce larger fruits and may encourage a tree that has been bearing on alternate years to bear annually.

If you do not prune extra spurs on the limbs, you may want to remove old ones. Spurs tend to produce fruit for several years, growing longer each year. After about three to five years, depending on the cultivar, the spur begins to age and lose vigor. Pruning off old spurs encourages new ones to grow.

SPUR PRUNING AND THINNING

You will develop an understanding of spur pruning and thinning as you watch your tree grow: When spurs are too crowded, too many fruits will form; when spurs are too old, buds decline in vitality. Try to thin spurs, either by removing them entirely or by cutting off buds, before they become too crowded. You can thin spurs in late winter or early spring each year.

ESPALIERS, CORDONS AND FANS

Each of these pruning and training techniques gives a somewhat formal look to your bush or tree. But these techniques may also increase the size and health of your fruit since you'll need to keep spurs pruned and branches trained to their supports, where they'll be certain to receive full light.

You must use a plant that bears on spurs to form an espalier, cordon or fan because you will be pruning off many laterals as well as branch tips when you prune and train your plant. Espaliered trees are pruned and trained so they grow, more or less, in two dimensions, or on a flat plane. Most homegrown espaliers are trained against a wall, generally because it is easier to attach wires firmly into a wall. Walls also store heat and provide some protection against inclement weather.

All branches on an espaliered tree are trained and tied to grow horizontally. When you first plant your tree, prune it to a whip. If there are already suitably placed branches, tie these to the wires and prune off all the rest. As the tree grows, retain only those branches that can be tied to the horizontal supports. You need to prune in late summer or early fall and again when the tree is dormant. Late in summer, cut back shoots as shown below. In winter, prune fruiting spurs as described on page 57. Also check to see if any of the shoots you pruned the previous fall have regrown. If so, head them back.

Cordons are pruned to a single stem and usually trained at a 45 degree angle. At planting, prune to a single whip or to a branch, if one is growing at just the right height (18 to 24 inches) and angle. Head back the leading shoot or branch and tie it to the support. Prune the current year's shoots as

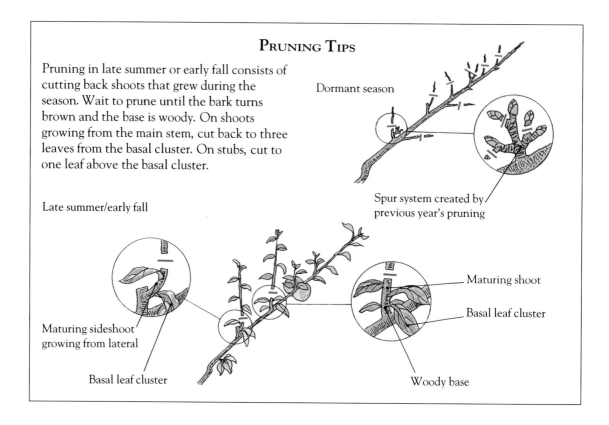

PRUNING TIPS

Pruning in late summer or early fall consists of cutting back shoots that grew during the season. Wait to prune until the bark turns brown and the base is woody. On shoots growing from the main stem, cut back to three leaves from the basal cluster. On stubs, cut to one leaf above the basal cluster.

Dormant season

Spur system created by previous year's pruning

Late summer/early fall

Maturing sideshoot growing from lateral

Basal leaf cluster

Maturing shoot

Basal leaf cluster

Woody base

▶ **Cordon trellises** *must be set up before you plant. You will need strong wires on top and bottom to form the support for the diagonal wire or thin wooden board that you tie the trunk to as it grows. Several trees or bushes can be cordoned in series.*

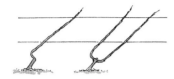

▶ **Espalier trellises** *are made by securing a series of horizontal wires to a wall, 12 to 18 inches apart. Make the trellis as high and as wide as you wish to allow the mature tree to grow.*

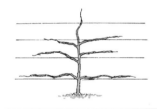

▶ **Fan-shaped trellises** *are stabilized by attaching top and bottom wires to posts or a wall. Place a stake in the center and wire thin wooden boards to the top and bottom wires, placing them to form a fan.*

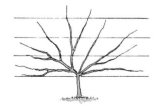

described for espaliers, and prune spurs in winter. Once the stem reaches the desired height, prune it back while it is dormant.

Fans are formed from branches trained to radiate from the central leader, with the first branch placed 18 to 24 inches above the soil. After four to six good fan branches have been selected, prune back the central leader. Maintain fans as you do espaliers, pruning in winter and late summer or early fall.

PRUNING AND TRAINING SHRUBS

Most bushes require only routine pruning and no training. However, because their needs vary, each is listed separately here, and you should check the "Fruit Directory," beginning on page 98, for specifics on each crop.

Currants, gooseberries and jostaberries: Though you can grow each of these bushes as a hedge, single-trunked standard, a "tree" or a freestanding bush, the bush form is by far the most common and easiest. After planting, cut back all the canes to two to four buds. Every year, early in the spring while the plants are still dormant but after the soil has thawed, cut out all but four or five of the old canes and allow four to five new canes to develop. By the third year, the bush will have about 16 canes and will have begun to bear well. Starting in the fourth spring, cut out the four-year-old growth, allowing new sprouts to take its place.

Black currants and jostaberries fruit on one-year-old wood, while red currants, white currants and gooseberries fruit most prolifically on spurs growing from two- and three-year-old canes. These pruning directions allow both types of plants to bear well. The important things to remember are to allow light to penetrate all parts of the bush and to remove canes that are four or more years old.

Gooseberry growers sometimes prune differently depending on whether the cultivar is upright or spreading. To promote good light penetration, the tips of upright cultivars are pruned to outward-facing buds. However, spreading cultivars are tip-pruned to inward-

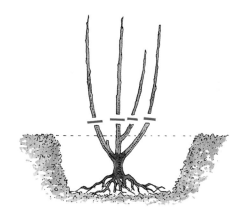

▲ **At planting,** *prune currants, gooseberries, and jostaberries by cutting back the canes to two to four buds.*

▲ **Routine pruning for bushes** *should be done in early spring while the plants are dormant. Prune out all but four or five of the oldest canes and allow the same number of new canes to develop. Starting in the fourth year, cut out the four-year old growth, letting new shoots take its place.*

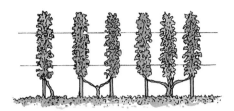

▲ **Cordon systems** *are suitable for red and white currants, jostaberries and gooseberries, since, unlike black currants, they all fruit on spurs. Set up the trellis as described for trees (see page 59) and make sure it is sturdy.*

facing buds to keep the plant within bounds. As long as you are pruning to allow light into the plant, you can choose either method.

If grown as bushes, none of these shrubs requires training. However, commercial growers often train them as cordons, particularly red currants and white currants. Light availability is never a problem with cordon systems, and it's easy to pick cordoned fruit.
Blueberries: Pruning can make all the difference between good-size fruit and fruit that is too small. When you plant highbush or half-high blueberries, prune off only the stray broken branch. Other than removing unhealthy growth, let them grow as they please for three years. In the fourth spring, cut off all the weak canes just above the soil surface and cut back weak laterals to strong, outward-facing buds. In the fifth spring and thereafter, cut out canes older than four years to let new growth develop. If berries diminish in size, it may be that plants are setting too much fruit. Rather than thinning the fruit, cut off the weak tips from some of the branches. To renovate old bushes, cut out half the canes, removing the other half the following year.

Rabbiteyes require even less pruning. No canes, other than weak growth, are removed until the bush becomes too dense for good exposure to light and air. Then cut out canes selectively to open up the bush. Rabbiteyes

bear on the tips of their branches, so only trim these if you need to decrease production in favor of getting larger fruit.

Lowbush blueberries are usually burned every two years in commercial production. However, this cultural technique is not advisable in a garden. Instead, cut off canes that have fruited, either in the fall or early the following spring, just above the crown. This allows new canes to form and lets light into the center of the plant.

PRUNING AND TRAINING BRAMBLES

Brambles include raspberries, blackberries, boysenberries, youngberries, tayberries, loganberries, wineberries and dewberries. Pruning depends on bearing time. Most brambles fruit on one-year-old canes. However, fall-bearing cultivars, sometimes called everbearing cultivars, fruit on canes produced in the current season as well as lower down on the previous year's canes and so they are treated differently from summer bearers.
Pruning summer-bearing cultivars: This is simply a matter of cutting out the canes that fruited, either in the fall after frost or early the following spring. This allows last year's primocanes sufficient light and adequate air circulation to fruit well during the coming summer. Fruit production is enhanced by

---

BRAMBLE TERMS

Before discussing pruning techniques, some bramble terminology needs to be explained.
Primocane usually refers to first-year canes.
Floricanes are canes on which blooms and berries grow.

Some authorities use the word "floricane" to indicate canes that have borne fruit and need pruning and "primocane" to indicate those that will bear fruit in the present year. To avoid confusion, pay careful attention to how the words are used.

pruning back the primocanes. When you cut away canes that have fruited, trim the canes that have not yet borne fruit to about 30 inches above ground level. They will respond by growing more blooming and bearing laterals.

Pruning fall-bearing cultivars: Fall bearers can be treated one of two ways. You can prune them as you would summer bearers, cutting out only the two-year-old canes and leaving the primocanes, or you can mow the whole patch several inches above ground level in the early spring. The latter method has the virtue of simplicity. It can also rid a planting of a great many disease problems. However, since these cultivars produce fruit on the old wood in July before making their large fall crop on new wood, it cuts yields.

Suckers: Brambles not only produce new growth from their crowns but they also produce suckers, or shoots, from their extensive root systems. You can use suckers to form a hedge if you space them widely. However, you can also pull or dig them out. In dense plantings, thinning out the suckers helps to keep air circulation high and the planting within reasonable bounds. If the suckers are strong and healthy, you can transplant them to a new area.

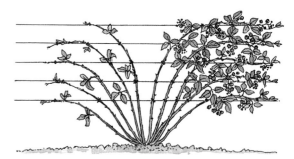

▲ **Multiple-stemmed cordons** *are practical for red and white currants, jostaberries and gooseberries. In small spaces, a single stem fits better, but if you have the room, train two or three branches to upright stakes or thin boards attached at the top and bottom to sturdy wires.*

Training brambles: Most of the numerous bramble training systems are designed to support bearing canes so they receive adequate air and light. The majority of home trellising systems depend on wires and posts. Of all the systems, the "T-bar" trellises with two sets of wires are the easiest to manage. There are two common ways of handling canes with such a system. In one, canes that will bear fruit are tied to the bottom wire, usually when tips are pruned. When these canes become long enough, they are tied along the top wire. The young canes that will

▶ **T-bar trellises** *are easy to make and work well for all nontrailing bramble cultivars. Use sturdy wood and attach the crossbars with screws rather than nails. Attach 9- or 10-gauge wire to the crossbars with staples. The top wires should be set at 5 feet, while the bottom wires are usually 2 feet above the soil surface. A space of 2 feet between the wires will accommodate a single row of pruned brambles without causing crowding.*

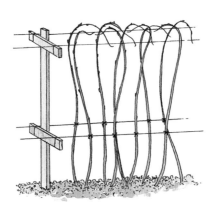

▲ **Drape canes** *over the wires when they reach the top. They will be easier to pick and prune afterward.*

form the following year's crop are left to lie on the ground. In fall or early spring, they are pruned back and tied to the bottom wire, and the process repeats itself.

Growers in humid climates sometimes prefer to get all the canes off the soil surface to try to diminish disease problems. In this case, they often separate the canes into those that will bear the current year and those which will bear the following year. They tie the bearing canes to one side of the plant and the nonbearing ones to the other. This makes both picking and pruning more convenient.

Strawberries: Strawberries require consistent pruning to remain productive. There are various options from which to choose. The double row is easy to maintain and gives high yields. The matted row, where you allow runners to set willy-nilly, is easy to maintain, but yields in the center of the bed are low and strawberries are small. Diseases can also proliferate in the crowded area. The Hill system gives large berries and protects against fungal diseases.

PRUNING AND TRAINING GRAPES

Four types of grapes are grown in the United States. In areas with mild winters such as California, European grapes are often grown.

American grapes like 'Concord' and 'Niagara' are grown in the North, muscadine grapes are grown in the Southeast, and American-European hybrids like 'Pinot Noir', because of their mixed parentage, are grown in many regions of the country.

Grapes require vigorous pruning. Left to their own devices, they will travel as much as 100 feet, generally bearing their fruit too high in a tree for easy picking. Unpruned grapevines also lose productivity. Grapes grow on one-year-old wood. If you leave old canes on the vine, the plant will spend too much energy sustaining the old growth and too little making fruit.

Pruning and trellising systems for grapevines must be discussed together because they

▲ **Strawberries: The double row system** *works with mother plants spaced 2 feet apart. Pinch off all but two runners that form on opposite sides of the plant, and set them 1 foot from the mother to form two new rows.*

▲ **Matted rows** *require the mother plants to be spaced 2 feet apart, in rows 3 feet apart. Allow runners to fill the areas in between, tilling to make paths if necessary.*

▲ **The Hill system** *has the mother plants spaced 1 foot apart in rows 3 feet apart. Pinch off all runners before they root. If you want more plants, set soil-filled pots under runners, as illustrated on page 68.*

are so interdependent. Here we look at two popular systems – the four-arm Kniffen system and the Munson system.

The four-arm Kniffen system: This system makes for easy pruning and high productivity.

Once planted, prune each vine to a single stem with two buds. Over the summer, these buds will develop into shoots. Tie these to the trellis wire to keep them off the ground.

The following spring, choose the strongest-looking shoot to turn into the trunk of the grapevine, and prune off all the other growth. Tie this cane to the trellis, either at the top wire, if it reaches, or to the bottom wire. Shoots will grow from this stem over the summer. As they do, pinch off any that form below the bottom wire. Choose four shoots, two on each side, growing close to the wires to keep. These will be the arms on which grapes will grow the following year. Prune off the other shoots, being careful in each case to leave two buds placed close to the wires. These are your "renewal buds," and shoots that grow from them will replace the wood you will cut at the end of the following year.

Meanwhile, new shoots will be growing. Prune off all but four, again placed so that they can be trained to the wires. Remember to leave renewal buds for the following year's growth and prune off the rest. Do not tie these new shoots to the wire until the old wood has finished fruiting and has been pruned off.

Grapes must sometimes be thinned. If your third-year vines set more than three or four clusters, grit your teeth and pinch them off while they are tiny. Older plants can produce 8 to 15 clusters on each arm without harm. If your plants try to outdo that, prune off the excess, remembering that you want to reduce crowding as well as numbers.

The Munson system: The Munson system is particularly good for vigorous muscadine grapes, which grow in the Southeast.

THE FOUR-ARM KNIFFEN SYSTEM

The four-arm Kniffen system is popular with most gardeners. Set up your trellis before you plant the vines. Make sure that the trellis is sturdy, since grapevines are very heavy and may grow and bear for a generation or two. All grape pruning and training systems can leave too many fruit clusters on the vines. If grapes are small, this may be the reason. Thin fruit clusters in subsequent years.

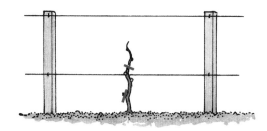

1 **After planting,** *prune the vine to two buds near the bottom wire. Prune back top growth above the wire and pinch off other buds.*

THE MUNSON SYSTEM

The Munson system raises the vine higher off the ground than the Kniffen and is less congested, thereby reducing the risk of fungal disease.

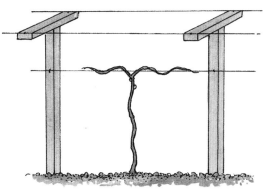

1 **Follow steps 1 and 2** *described above. In spring, choose two shoots to tie along the lower wire. Prune off all others, leaving renewal buds for next year.*

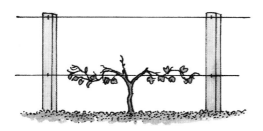

2 **Shoots will develop** *from the two buds over the summer. Tie them to the wire.*

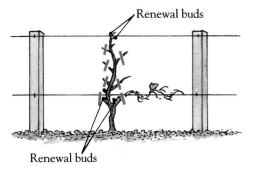

3 **In spring**, *choose the strongest shoot to become the trunk. Prune off the laterals, leaving two renewal buds near each wire.*

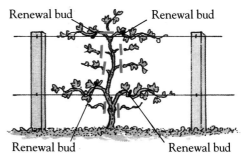

4 **Over the summer**, *pinch off all of the shoots, leaving four growing close to the wires. Leave renewal buds close to the wires.*

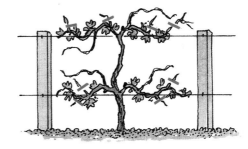

5 **Prune all new shoots** *over the summer except those for next year's fruit. Tie them in place at the end of the season.*

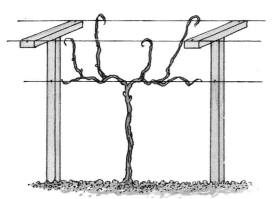

2 **As shoots grow** *from the arms the following year, drape them over the top wires. Laterals from the draped arms bear fruit the following year.*

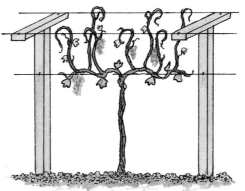

3 **Allow two new shoots** *to form each year for the following season.*

Harvesting

Harvesting is the most exciting part of growing fruit. Few gardening experiences rival that of loading up a garden cart with empty baskets on a sunny afternoon and heading off to harvest a crop of luscious fruits.

Y ou may be surprised at the yields from your fruit plants. Most people discover that they have to dust off some old preserving techniques and learn some new ones to keep up with the bounty from even a small planting. Others learn, much to their delight, that friends and neighbors react very differently to a basket of strawberries than they do to offerings of extra tomatoes and zucchini.

Harvesting is straightforward. Because vitamin C content is highest in the afternoon, try to pick then. Almost all fruits bruise easily, so handle them carefully and don't pile them so

HOW TO USE YOUR FRUITS

Fruit	Uses	Notes	Fruit	Uses	Notes
Apple	Fresh, baked goods, sauces, jelly, juice, preserves, dried	Select appropriate uses for each cultivar you grow. Store only unbruised firm fruit.	Cherry, sour	Fresh (some cultivars), baked goods, sauces, jam, preserves, canned in sugar syrup, frozen, dried	Some people like the flavor of sour cherries fresh. Most people prefer sour cherries to sweet in pies, jams and preserves.
Apricot	Fresh, dried, preserves, baked goods, canned in sugar syrup	Handle carefully when picking; apricots bruise easily.	Cherry, sweet	Fresh, baked goods, sauces, jam, preserves, canned in sugar syrup, frozen, dried	Adjust sugar if using sweet cherries in recipes meant for sour cherries. Use dried cherries in baked goods.
Blueberry	Fresh, baked goods, jam, ice cream, pancake syrup, frozen, canned in sugar syrup	One of the most versatile fruits. Do not wash until using. Do not pile deeply in picking baskets.	Citrus: lemon, lime, grapefruit, orange	Fresh, juices, marmalade, baked goods	Learn ripening characteristics of your cultivars. Use clippers to cut from tree.
Brambles: blackberry, boysenberry, dewberry, loganberry, raspberry, tayberry, youngberry	Fresh, baked goods, frozen plain, frozen in syrup, vinegar, pancake syrup, jam, ice cream	Do not wash until using. Be careful not to crush in picking baskets.	Currants: black, red, white	Fresh, dried, baked goods, jelly, jam, frozen	Pick in afternoon for highest vitamin C content.
			Elderberry	Baked goods, jellies, juices, sauces	Do not wash. Be careful not to crush in picking baskets.

high in a basket that they crush each other.

Some tools are useful; long-handled fruit pickers can extend your reach enough so you don't need a ladder. A good pair of clippers is helpful for cutting grapes, currants and gooseberries, and if you have many blue-berries, you might consider buying a blueberry rake from a commercial supply house.

Ripeness is sometimes hard to determine without actually tasting the fruit. By all means, start "testing" when you think the fruit should be ready. However, you will soon learn exactly what coloration and "feel" indicates that it's picking time.

▲ **Succulent homegrown peaches** *taste as different from store-bought as homegrown tomatoes do from winter imitations. You won't regret planting this tree!*

Fruit	Uses	Notes	Fruit	Uses	Notes
Fig	Fresh, dried, preserves, baked goods	Fresh ripe figs are a treat most northerners have yet to taste.	Pear, Asian	Fresh, poached, baked goods, preserves	Pick Asian pears when fully ripe. Handle carefully to avoid bruising.
Gooseberry	Fresh (some cultivars), pies, preserves, frozen	Pink cultivars are sweet enough to eat fresh.	Pear, European	Fresh, canned in sugar syrup, baked goods, preserves, poached	Pick before fully ripened, when you can pull from tree. Some cultivars store well in a root cellar.
Grape	Fresh, jelly, juice, frozen puree, dried	Frozen puree can be used to make a pancake syrup.			
Kiwi	Fresh, tarts, jelly	Kiwi jelly is easy to make and beautiful to look at.	Persimmon	Fresh, puddings, baked goods, frozen whole	Wait to eat fresh until fruit is very soft. Firm fruit is sour.
Nectarine	Fresh, baked goods, canned in sugar syrup, jam	Harvest when you can pull fruit from tree. Ripen inside if necessary.	Plum	Fresh, baked goods, sauce, jam, jelly, dried (prune plums)	Use slightly underripe plums for cooking. Handle carefully to avoid bruising.
Pawpaw	Fresh, custard, puddings, preserves, dried	Pick when fruit softens slightly. Ripen inside for best flavor.	Quince	Jelly, sauce, preserves, canned in sugar syrup	Pick when fragrant. Handle carefully to avoid bruising.
Peach	Fresh, baked goods, canned in sugar syrup, jam, preserves	Some cultivars ripen best indoors. Pick when fully colored and soft around stem.	Strawberries	Fresh, jam, pancake syrup, baked goods, frozen	Do not wash before using. Handle carefully, as fruits spoil when bruised.

Propagation

Propagating fruit plants adds to the enjoyment of growing them. You can have more of a choice plant at no charge by being patient and using some simple techniques. Some plants, such as brambles, propagate themselves so easily that all you have to do is dig up and separate new ones. But others, such as apple cultivars, require you to learn specific skills. The following sequences show common and relatively easy techniques for you to try.

Most fruit plants are propagated vegetatively rather than by seed. Vegetative reproduction depends upon rooting some part of the plant, such as a stem or branch tip, dividing an established plant crown so that each part includes both roots and a stem, or joining a bud or branch to a selected rootstock so the two grow together into one plant.

Some fruits, such as blackberries, naturally reproduce vegetatively, growing new plants from suckers on their roots. Others, such as apples and cherries, are more likely to grow from seeds. While a chance seedling can be a pleasant surprise — some of our favorite cultivars originated this way — the majority of these seedling plants will produce inferior fruit or have roots that are inappropriate to your climate. Vegetative propagation, on the other hand, produces plants that are just like their parents and bear fruit earlier than seed-grown plants.

Vegetative propagation techniques include root divisions, rooting both hardwood and softwood cuttings, tip and stem layering, and grafting. While these techniques may seem intimidating at first, they are simple to learn and relatively easy.

Root divisions are difficult and best left to

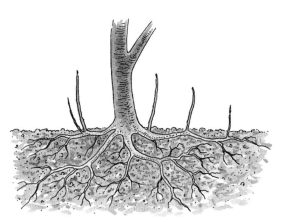

▲ **Suckers** *from a healthy plant or rootstock can be severed from the parent, moved to a new location or a pot where they can establish themselves, and then grown on or used as the rootstock for a chosen cultivar.*

▲ **Strawberry runners** *root easily. If your patch is crowded but the plants are healthy, place selected runners over soil-filled pots. After they root, transplant them to a new bed. Do this early enough so that the plants can get established before frost.*

professionals. But propagating from suckers is not much different than dividing an over-grown patch of bee balm, since it involves digging up a sucker formed away from the row, making sure you get enough roots for it to transplant well.

Taking cuttings from fruit plants is no different or harder than taking them from houseplants. Layering, whether of branch tips or stems, couldn't be easier. However, grafting techniques will probably be new to you. As long as you follow the grafting guidelines given on page 70, you will soon discover that grafting is really no more

difficult than starting plants from seed. Work slowly and carefully at first, be certain to sharpen your propagating knife and use a small magnifying glass if your eyes don't focus well on close work.

GRAFTING

Grafting is both useful and fun. If you have space for only one apple tree, for example, but want to grow a cultivar that requires a pollinator, you can solve the problem by grafting several scions of an appropriate pollinator onto your tree. Similarly, you may want to try an heirloom fruit. Some heirlooms

TIP LAYERING

1 **Tip layering** *is the easiest and most effective way to propagate brambles such as raspberries. Begin by pinching the growing tip of a new basal shoot early in the spring.*

2 **After new tips have formed,** *gently pull the stem to the ground and make a small trench or depression to lay it in. Hold it in place with a wire or plastic staple.*

3 **Bury the tip,** *including the leaves, firm the soil and water well. The buried stem will swell and grow roots over the summer. Remember to keep the buried stem well watered in dry conditions.*

4 **New leaves will show** *late in the season. When they do, you can cut the mother stem from the buried portion. Wait until the following spring to dig and move the new plant.*

attract every pest and disease known to apples, but others like 'Summer Rambo' and 'Yellow Transparent' are not only delicious but they are also vigorous and disease-resistant. If you know how to graft, you can add them to appropriate trees or rootstocks (plant bases).

Grafting guidelines: For successful grafting, follow the guidelines listed below.

- Check with your local Extension agent or a fruit growers' organization to be sure that the rootstock and scion (stem) you plan to graft are compatible. (You can also contact the North American Fruit Explorers, known as NAFEX, at Route 1, Box 84, Chapin, IL 62628.)
- Time your grafting correctly. Whip grafts are made in early spring, just before or at the time the buds begin to swell. The correct time for budding is late summer.
- Collect scionwood when it is dormant, wrap it in damp peat moss and store at temperatures just above freezing until you are ready to use it.
- Cut the scion and the stock so they fit together smoothly at the cambium layers. Coat them with grafting compound to protect the union from drying out.

WHIP-AND-TONGUE GRAFTING

The technique of whip-and-tongue grafting is used to join a scion to a rootstock.

1 **Cut** *the bottom of the scion and the top of the rootstock at the same angle. The cut should be from 1½ to 2½ inches long, depending on the diameter of the wood.*

2 **Make angled slits** *in each cut, forming the "tongues" that will fit together. Slide the rootstock tongue into the scion slit and secure it with rubber bands or twine.*

SPLICE GRAFTING

Simpler than whip-and-tongue grafting, splice grafting does not require precisely cut slits.

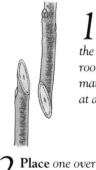

1 **Be certain** *that the diameters of the scion wood and rootstock wood match, and cut them at a uniform angle.*

2 **Place** *one over the other.*

3 **Use rubber bands** *or waxed twine to hold them in place while the cuts knit together with new growth. Leave the wrapping in place until the graft has taken, then remove it carefully.*

T-BUDDING

The technique of T-budding is used to graft stone fruits, citrus and vines. Success depends on matching the cambiums of stock and bud exactly. Bud in June in long-season areas and early fall in short-season regions.

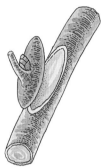  

1 **Begin by** *taking a bud from the current year's growth. Slice deeply enough to get some wood under the bud, too.*

2 **Make a T-shaped incision** *in the stock where you want the new branch to grow, then gently lift the bark on both sides of the cut.*

3 **Slip the bud** *into the T-cut.*

4 **Now wrap** *the union well with waxed twine. In 3–4 days, cut off the rootstock top, leaving one or two leaves above the bud.*

5 **Wait several** *weeks until the bud has taken, then remove the waxed twine.*

6 **Cut the** *rootstock top back to the bud when it has grown into a shoot 6 to 8 inches long. Remove all other shoots and buds from the stock.*

TWO KINDS OF CUTTINGS

▲ **Softwood cuttings** *are taken from the tips of new growth. Make the cuttings 4 to 6 inches long (left). Remove all leaves except one or two at the top (right). Bury the cuttings in moist peat-perlite mix and set them in low light until they root.*

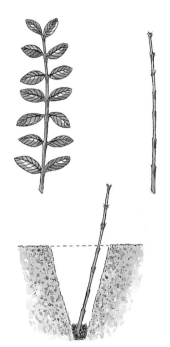

▲ **Hardwood cuttings** *are taken in the fall. Cut an entire shoot (top left). Make a straight cut on the bottom and a sloping cut on the top, and remove all the leaves (top right). Insert the stem so that all but three buds are buried, above. Leave over winter and transplant in the early spring.*

Chapter

3

COPING WITH PESTS AND DISEASES

PESTS AND DISEASES are a fact of life when growing fruits and berries. Along with a fear of pruning, they are one of the major reasons why more backyard gardeners don't grow fruit. But a commonsense approach to pest and disease control will put you in charge of any problems. And, just to put things in perspective, pests and diseases don't keep gardeners from growing vegetables! In this chapter, you'll learn the importance of preventive practices like good cultural techniques, careful monitoring and choosing resistant species and cultivars when available. You'll get a rundown of major fruit pests (including animals) and diseases and the problems they cause. And you'll learn how to prevent or control these problems with traps and barriers, biological controls, botanical sprays and mineral sprays and dusts. Other effective tricks and techniques are included throughout the chapter.

Cultural Techniques

How you manage your fruit plants—as well as the other plants and structural elements of your backyard—can make all the difference to plant health. Aim to develop cultural techniques that encourage natural predators and parasites, protect plants from adverse environmental effects and create conditions for good health and productivity. Good cultural techniques start with choosing plants that have been bred to resist pests and diseases. Planting resistant cultivars is the easiest way to give your garden a head start. For example, 'Liberty' apples are resistant to four major diseases. You'll find other cultivars in the "Fruit Directory," beginning on page 98.

It is important to carry out routine maintenance chores at the right time. For example, fertilizing later than midsummer, even with a slow-release amendment such as compost, will encourage late, and very tender, growth that is vulnerable to winter damage. Timing can be critical to the health—or even the life—of your plants.

Each time you add a new plant to the landscape, you change your yard's dynamics. Ask yourself what effect will it have on light availability and shade patterns, air circulation, relative humidity levels for surrounding plants and indigenous populations of birds and insects.

Think about the advantages and drawbacks of all cultural techniques before you use them. For example, in a rainy spring and early summer, a deep mulch on your strawberry bed can minimize the rotting diseases that infect berries lying on the soil surface. However, it can also encourage both mice and slugs. If disease is a greater problem, your choice is clear. However, you should also know that you may need to be more than usually vigilant about pest damage.

If problems with pests and diseases do occur, think through the conditions that promote these problems and try to develop practices that could reverse or correct them. For example, if apples have high levels of scab infection, you need to prune off and destroy infected leaves and fruit as well as gather and dispose of all drops.

Traps, tricks and barriers are just some of the cultural techniques you can use to protect your fruit plants. But while these techniques may be optional in a vegetable garden, they are mandatory for pest control in fruit crops.

▲ **Keep a beehive or two** *in your fruit garden, since bees are some of the most effective pollinators. The bees will work on your plants before they fly in search of other flowers, and the honey you collect will be a bonus.*

TRAPS, TRICKS AND BARRIERS

Each time you encounter a pest, think about ways to minimize its damage so you don't have to spray. Traps, tricks and barriers are all effective, particularly on a home garden scale. For example, some commercially available traps are meant to monitor pest populations so farmers can decrease spraying frequency. But in a small planting, these same traps may be able to catch most of your pests. The ideal insect trap is very specific in its action. Many companies sell traps and other insect-controlling devices. Look for traps that are designed to attract, confuse, repel or barricade the particular pests on your fruit plants.

▼ **Sticky bands** *are wrapped around tree trunks to trap adult insects and caterpillars. Tie the bands tightly to the tree to prevent pests from crawling under them, then reapply sticky material as needed.*

◀ **Delta traps,** *with sticky bottoms treated with pheromones or other attractants, are used to capture both large and small adult moths. Specific lures are sold separately for these traps as well as for codling-moth traps.*

▲ **Sticky balls** *trap many insect pests, including apple maggot flies and plum curculios. Hang them from June till September and wipe and reapply sticky substance as needed.*

▶ **Deodorant soap** *may seem like a strange decoration for your fruit trees, but it repels deer effectively. Leave the wrappers on the bars and hang them 4 to 6 feet apart. Use soap in combination with net bags of dirty human hair for maximum effect!*

▲ **Hose pieces** *may not look like dangerous snakes to you, but to a bird, who recognizes shapes, they do. To be certain that the pieces remain effective deterrents, move them around the planting every day or so.*

Natural Troubleshooters

Botanical and biological pesticides are not nearly as dangerous as synthetically derived poisons because they break down faster. Nonetheless, some botanical pesticides are highly toxic for a short time after they're applied, so it's wise to develop careful habits if you use them.

Botanical and biological pesticides are made from plants or natural substances. Once applied and exposed to sunlight and air, they quickly break down into materials that "wouldn't hurt a fly"—or anything else. But before they decompose, they can be very effective.

Using pesticides, even botanicals and biologicals, is a last-resort measure. Many of these materials, like milky disease and *Bacillus thuringiensis* (BT), are fairly specific in their action; they kill insects in one family while leaving members of another group unharmed. But others termed "broad-spectrum pesticides," like rotenone, kill members of a wide range of insect families, including some of the beneficials in your garden. Before resorting to a pesticide, use preventive measures such as traps and barriers, as well as biological controls such as beneficial insects. But if none of these methods is effective, applying one of the pesticides listed in "Common Organic Pesticides" on the opposite page may be necessary.

PESTICIDE POLICY

- Research the pests on your plants to discover cultural methods or biological controls that are effective.
- Store pesticides where they can't be reached by children, pets or wild animals.
- Mix only the amount you need that day to avoid a disposal problem.
- Wear protective gear—rubber gloves, a face mask and/or goggles and a rain suit—when appropriate (see "Common Organic Pesticides" on the opposite page).
- Use an appropriately sized sprayer or duster to apply materials.
- Spray at the correct time of day—early morning in most situations.
- Avoid drift by spraying when the air is calm.
- Cover both sides of leaves with spray materials.
- Spray only as often as necessary and only on affected plants.

MAKE YOUR OWN SOAP SPRAY

You can use household soaps such as Ivory Snow to make your own insecticidal soap solution.

Be aware that many products commonly known as soap also contain impurities, such as perfumes and whiteners, that can damage plants. For pest control, it is important to use pure soap; avoid detergents or soaps with additives.

To prepare a homemade soap solution, mix from 1 teaspoon to several tablespoons of soap per gallon of water. Start at the lower concentration and adjust the strength to maximize pest control while avoiding plant damage.

From *The Organic Gardener's Handbook of Natural Insect and Disease Control* (Rodale Press, 1992)

COMMON ORGANIC PESTICIDES

Material	Pests Controlled	Notes	Material	Pests Controlled	Notes
Pest-specific biological pesticides			Rotenone	Aphids, beetles	From roots of legumes in the genus *Loncho-carpus*. Broad-spectrum pesticide. Some-what toxic to mammals; toxic to birds, fish and beneficials. Avoid using. Wear protective gear. Biological.
Bacillus thuringiensis var. *kurstaki* (BTK)	Caterpillars	Use immediately after mixing. Repeat at 7 to 10 day intervals. Biological.			
Milky disease, also called milky spore (*Bacillus popilliae*)	Japanese beetle grubs	Works best in warm climates, Zone 6 and above. Biological.			
Parasitic nematodes	Japanese beetle grubs, many soil-dwelling organisms	Works in warm and cool climates. Reapply annually. Biological.	*Ryania*	Caterpillars, Japanese beetles, citrus thrips	From the shrub *Ryania Speciosa*. Wear protective gear. For codling moth control, spray every 10 to 14 days, starting at petal fall. Mix with sugar to kill citrus thrips and use in hot, sunny weather. Biological.
Broad-spectrum (botanical and other) pesticides					
Citrus oils	Aphids, mites, caterpillars	From citrus peels. Commercial products may contain insecticidal soap. Botanical.			
Dormant oil	San Jose scale, oystershell scale	Spray before bud break.			
False hellebore	Caterpillars, sawflies	Wear gloves and protective gear. Botanical.	Sabadilla	Aphids, tarnished plant bugs, thrips	From seeds of *Schoenocaulon officinale*. Toxic to honeybees; moderately toxic to mammals. Wear protective gear. Spray every seven days to protect strawberries from tarnished plant bugs.
Nicotine	Aphids, scales, spider mites	From tobacco leaves. Wear a face mask when applying. Mix with soap for added efficacy. Highly toxic to all insects and mammals. Avoid using if possible. Botanical.			
Pyrethrin	Aphids, codling moths, spider mites, thrips	From pyrethrum daisies. Kills lady beetles. Also produces allergic reactions in some people. Wear protective gear. Biological.	Insecticidal soap	Aphids, mites, scales, thrips, mealybugs	Limit spray area, since soap kills many beneficials.

Insect Friends

Beneficial insects kill and parasitize amazing numbers of insect and mite pests every day. For the most part, they do this so invisibly that you are not even aware of the help they are giving your plants.

To keep populations of beneficial insects high, you need to follow a few simple guidelines. Most important is to use pesticides only when absolutely necessary. Many pesticides kill beneficials as easily as they kill pests.

Encourage beneficials to visit by growing a succession of small-flowered plants such as mints (*Mentha* spp.), sweet alyssum (*Lobularia maritima*), annual scabiosa (*Scabiosa* spp.), common yarrow (*Achillea millefolium*) and dill

Beneficial Insects and Their Prey

Beneficial	Pest Controlled	Beneficial	Pest Controlled	Beneficial	Pest Controlled
Beetles *Delphastis* spp.	Whiteflies, including sweet potato whitefly	**Flies** *Aphidoletes aphidimyza* •	Aphids	**True bugs** Assassin bugs	Caterpillars, flies
Ground beetles	Cutworms, slugs, snails, tent caterpillars, gypsy moths	Syrphid flies	Aphids	Minute pirate bugs •	Caterpillars, leafhopper nymphs, spider mites, thrips
Lady beetles •	Aphids, mealybugs, soft scales, spider mites	Tachinid flies	Caterpillars, Japanese beetles, sawflies	Spined soldier bugs •	Fall armyworms, tent caterpillars, sawfly larvae
Rove beetles	Aphids, nematodes, fly eggs and maggots				
Soldier beetles	Aphids, caterpillars				

(*Anethum graveolens*) close to your fruit plants. They will provide food for the beneficial insects through the growing season. In drought periods, leave a pan of water in the garden. Place several rocks in it to give the beneficials a place to stand while drinking.

PURCHASED BENEFICIALS

Beneficials take some time to build up populations in a new garden. While the garden is beginning to "come alive," one pest or another might build intolerably high populations and do real damage to your plants. Fortunately, you can purchase indigenous beneficials. Not only do these save your plants but they also allow you to avoid using a pesticide that might severely knock back the growing populations of beneficials.

Some beneficials may not be capable of overwintering in your climate. Buying a supply every year makes more sense than buying a bottle full of poison. Suppliers of beneficial insects are listed in the "Resource Directory," beginning on page 155.

The chart below lists beneficial insects as well as the insects or mites they control. Those that are marked with an asterisk are commercially available.

Beneficial	Pest Controlled	Beneficial	Pest Controlled	Beneficial	Pest Controlled
Lacewings Brown lacewing	Aphids, some caterpillars, mealybugs, mites, some scales, thrips	**Wasps** Braconid wasps*	Aphids, armyworms, codling moths, gypsy moths	**Arachnids** Predatory mites*	Spider mites, two-spotted spider mites
		Chalcid wasps 	Aphids, some caterpillars	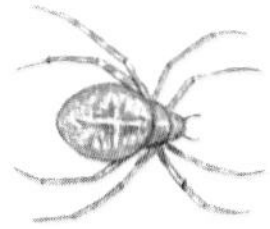	
Green lacewing* 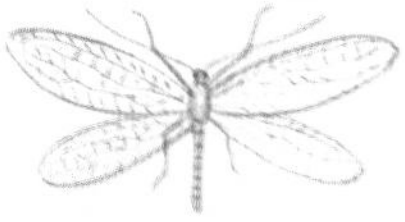	Aphids, some caterpillars, mealybugs, mites, some scales, thrips	Ichneumonid wasps 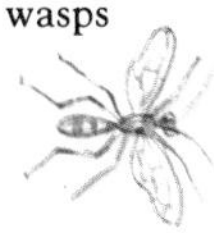	Caterpillars, sawfly larvae	Spiders	Many pests
		Trichogramma wasps* 	Aphids, other soft-bodied insects		
		Yellow jacket wasps 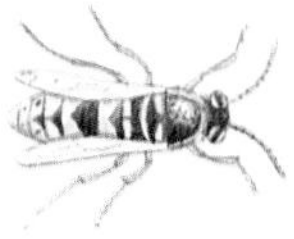	Caterpillars, flies		*This beneficial is available commercially.

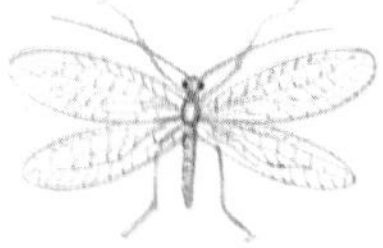

Fruit Foes

Insect pests can ruin what would otherwise be a good crop. To grow good quality fruit, you'll have to pay attention to pest control and management.

Rely on basic principles to manage fruit pests: be watchful all the time. Look under leaves and on the bark to find pests. Check to see that all growth is normal, and suspect a problem if you see leaf or fruit distortions. Practice preventive techniques.

TREES

APPLES

Codling moths: Pinkish, brown-headed larvae eat developing fruit. In late winter, spray with dormant oil and hang 1–2 pheromone traps per tree. Wrap trunk with sticky bands to trap larvae. In early spring, scrape loose bark off trunk to expose cocoons. In severe cases, spray ryania when most petals have dropped; repeat 3–4 times.
Also attack: Apricots, pears, quince

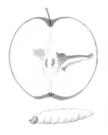

Plum curculios: Larvae cause fruit to drop in June or to rot or develop irregularly. From early spring, spread a sheet under tree and tap branches with a padded stick twice a day. Shake sheet over a bucket of soapy water to destroy weevils. Use red ball traps painted green and covered with sticky coating. Pick up all drops.

Apple maggot flies: Larvae bore through fruit, often causing it to rot or drop early. Pick up and destroy dropped fruit. In mid-June, hang red sticky balls at eye level, 3 feet from branch tips; 1–2 per dwarf tree, 4–8 per standard. Clean balls and reapply sticky coating every few days at first.

CHERRIES, SWEET

Cherry fruit fly maggots: Larvae feed on interior of fruit, causing shrunken or dropped fruit. Trap adults with red sticky balls hung in tree from midspring to midsummer. As a last resort, spray with rotenone. Scratching chickens may eliminate pupae.
Also attack: Sour cherries

Pear slugs: Larvae of pear sawfly skeletonize leaves. Handpick larvae from leaves. Spray insecticidal soap for large populations. Dust leaves with diatomaceous earth; this kills larvae that crawl across it.
Also attack: Sour cherries

TREES

CHERRIES, SOUR

Black cherry aphids: Weaken tree by sucking cell sap from leaves. Excreted honeydew can cause sooty mold to cover leaves. Encourage beneficials by growing small-flowered plants like dill nearby. Apply dormant oil in late winter before bud break. Use rotenone as a last resort.

FIGS

Ants: Crawl into fruit and eat interior. Place ant traps baited with boric acid at base of plant. Spread wood ashes around base of plant to keep ants away. Wrap trunk with sticky bands to trap ants.

Spider mites: Tiny mites suck cell sap, causing yellow stippling and dropped leaves. Leaves may be covered with fine webbing. Spray with dormant oil in very early spring. Spray tree with water daily. Spray insecticidal soap or pyrethrin. Release predatory mites. Use rotenone as a last resort, but not with predators.

NECTARINES

Oriental fruit moths: Larvae tunnel into twig tips in spring and into developing fruit in summer. Cultivate around tree before flowering to destroy overwintering larvae. Spray with BTK *Bacillus thuringiensis* var. *kurstaki*) every 2–3 weeks from early spring to fall. Use pheromone traps. Release parasitic wasp *Macrocentrus ancylivorus*. Spray superior oil to kill eggs and larvae.
Also attack: Peaches, most fruit trees

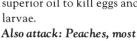

Peachtree borers: Grubs bore into trunk near soil surface and can girdle young trees. Look for gummy sawdust around holes. Choose resistant cultivars. Cultivate soil around tree to kill pupating moths. Insert flexible wire into each hole to kill larvae. In the worst cases, carefully dig out borers with a knife.
Also attack: Peaches

PEACHES

Scales: Suck sap. Leaves and branches may be encrusted. Leaves yellow; branches may die. Excreted honeydew attracts sooty mold. Spray dormant oil in late winter. Spray superior oil. Wrap trunk with sticky bands to trap ants, which encourage scales (see page 75). Use pyrethrin or rotenone as a last resort.
Also attack: Persimmons

TREES

PEACHES

European red mites:
Weaken tree by sucking sap
from leaves. Spray tree
weekly with plain water.
Spray in summer with
insecticidal soap. Spray
superior oil when leaf buds
swell in spring. Release
predatory mites.

PEARS

Pear psylla: While sucking sap,
psylla transmits pear decline, a
mycoplasma disease. Excreted
honeydew attracts sooty mold.
Choose a resistant rootstock.
Spray with dormant oil in late
winter. Spray with superior oil
or insecticidal soap 3–4 times
during summer.

PEARS, ASIAN

Aphids: Suck sap, weakening
plants. They can also transmit
fire blight spores. Excreted
honeydew attracts sooty mold.
Spray with dormant oil in late
winter. Spray with insecticidal
soap in early spring. Encourage
beneficials by growing small-
flowered plants like dill nearby.

Pear slugs: Larvae of pear
sawfly skeletonize leaves.
Handpick and destroy larvae
in soapy water. Spray
insecticidal soap or, if you
can get good coverage, dust
tree with diatomaceous earth.

PLUMS

Aphids: Weaken tree by sucking
cell sap. Excreted honeydew can
promote sooty mold on leaves,
twigs and fruit. Spray dormant
oil in late winter to kill eggs.
Spray insecticidal soap when
leaves enlarge in spring. Grow
small-flowered plants nearby to
attract beneficials. Use
pyrethrin or rotenone as
a last resort.

Plum curculios: Larvae burrow
into fruit, causing it to rot, drop
or become misshapen. Knock
weevils from tree as described for
apples (see page 80). Spray
severe infestations with a mix of
pyrethrin, ryania and rotenone;
repeat in 7–10 days. Collect and
destroy all dropped fruit. If
possible, let poultry range
under trees.
***Also attacks: Apricots, sour
cherries, peaches***

Red spider mites: Weaken tree
by sucking cell sap. Leaves may
show webbing and yellow
stippling. Spray dormant oil in
late winter or very early spring.
Spray superior oil. Spray
insecticidal soap or pyrethrin.
Identify species and release
appropriate biological control.

QUINCES

Oriental fruit moths: Larvae
bore into fruit and twigs.
Cultivate around tree to a
depth of 4 inches in early
spring to expose over-
wintering larvae. Spray
superior oil in spring and
early summer. Apply
pheromone patches to one
out of four trees.

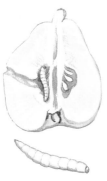

FRUIT-BEARING BUSHES

BLUEBERRIES
Blueberry maggots: Small
(3/8 inch) maggots eat fruit from
inside. Disrupt life cycle by
removing and destroying all
softened, infested fruit. Trap
adult flies on sticky red balls
hung on the bushes before the
berries turn blue.

Cherry fruitworms: Worms
feed on berries and web them
together. Pick and destroy
infested fruit to disrupt life
cycle of fruitworm. As soon as
flowering is over, spray bushes
with rotenone.

CURRANTS, GOOSEBERRIES AND JOSTABERRIES
Currant borers: Larvae eat
interior of canes, causing them
to weaken, break easily and
eventually die. Look for small
holes with specks of sawdust
around them. Cut out and
destroy affected canes. Cover
bushes with floating row covers
from midspring to midsummer, or apply
a summer oil spray to kill eggs.

Currant fruit flies: Larvae eat
berries from inside, causing
them to drop. Destroy all
infested berries as soon as you
find them. Rake area around
bushes to gather up any you
might have missed and mulch
with clean material. Plant early-
bearing cultivars to avoid the pest.

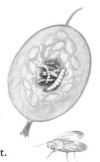

ELDERBERRIES
Elder shoot borers: Larvae
tunnel into branches to feed
on the wood. Shoots turn
yellow and die. Prune out
and destroy infested shoots.

STRAWBERRIES
Spider mites: Tiny mites
weaken plant by sucking sap,
causing yellow mottling on
leaves. Leaves and stems may be
covered with fine webbing.
Maintain even soil moisture
and spray plant with water daily.
Spray with insecticidal soap every
7–10 days or release predatory mites;
in this case, do not use soap sprays.

**Strawberry root weevils
(strawberry clippers):** Weevils
clip off edges of leaves and cut
off fruit at stems. Larvae bore
into crowns and roots, stunting
growth and killing plant.
Drench soil with parasitic
nematodes in early May or as
soon as weevils or their damage appear.
Cover plants with floating row covers in early
spring. Dusting with rotenone will kill weevils.

Tarnished plant bugs: Toxic
saliva of bugs causes bud drop and
irregularly shaped fruit. Cover
plant with floating row covers
from earliest spring through
harvest. After harvest renovate
June-bearers (see page 152), and
rake out and compost old leaves
and mulch from day-neutral cultivars.
Use sabadilla as a last resort.

BRAMBLES

BLACKBERRIES

Aphids: Suck plant juices, causing leaf curl; they may transmit viral diseases. Cover plant with a floating row cover in early spring. Grow small-flowered plants like dill nearby to encourage beneficials. Spray plant with insecticidal soap at first sign of aphids. Dig up and destroy plants exhibiting viral diseases.
Also attack: Boysenberries, dewberries, loganberries, raspberries, tayberries

Cane borers, raspberry red-necked: Cause cane tips to wilt and die. Dig up and destroy any canes with borer damage to interrupt pest's life cycle. Eliminate any wild brambles growing nearby.
Also attack: Boysenberries, dewberries, loganberries, raspberries, tayberries

Spider mites: Weaken plant by feeding on sap, causing leaves to turn yellow and drop; webbing may be visible on underside of leaves. Spray plant with water. Mulch to keep soil moist. Spray with dormant oil in early spring. After plants have leafed out, spray canes with insecticidal soap or introduce predatory mites.
Also attack: Boysenberries, dewberries, loganberries, raspberries, tayberries

LOGANBERRIES

Raspberry sawflies: Small (¼–½ inch) larvae feed on underside of leaves, skeletonizing them. Handpick larvae. Spray with BTK (*Bacillus thuringiensis* var. *kurstaki*) when larvae are small.
Also attack: Raspberries, tayberries

DEWBERRIES

Japanese beetles: Beetles skeletonize leaves, weakening the plant. Plants may be defoliated by heavy infestations. Handpick beetles. Apply milky disease (*Bacillus popilliae*) to soil around planting in fall or spring to kill grubs. Spray infested plants with ryania or rotenone.
Also attack: Raspberries, tayberries

Raspberry fruitworms: Adults feed on flowers and leaves. Larvae eat fruit from inside. Collect and destroy infested fruit during summer to interrupt life cycle. In early spring, spray with rotenone when flower buds swell and again when blossoms open.
Also attack: Loganberries, raspberries, tayberries

VINES

AMERICAN AND MUSCADINE GRAPES

Grape leafhoppers: Suck juices from leaves. Affected leaves show light spotting. Pest can transmit the virus that causes Pierce's disease. Encourage parasitic wasps by growing plants like dill nearby. Spray with insecticidal soap. In the worst cases, spray with pyrethrin/rotenone mix.
Also attack: European and hybrid grapes

Grape mealybugs: Suck juices from grapes and stems, excreting honeydew on which sooty mold grows. Grapes drop prematurely if stems are attacked. Touch bugs with a cotton swab dipped in alcohol. Introduce mealybug destroyer (*Cryptolaemus montrouzieri*) to deal with large infestations or spray with insecticidal soap.
Also attack: European and hybrid grapes

Grape berry moths: Cause webbing between grapes, holes in grapes and on leaf margins, and rolled leaf edges; also eat flower buds. Handpick larvae, checking leaf margins for pupae. For large infestations, spray with BTK every few days until control is achieved.
Also attack: European and hybrid grapes

Japanese beetles: Skeletonize leaves, weakening the plant. Cultivate around vines. Handpick beetles. Sprinkle milky disease or water in predaceous nematodes to kill grubs. Spray infested plants with ryania or rotenone.
Also attack: European and hybrid grapes

EUROPEAN AND HYBRID GRAPES

Grape cane gallmakers: Form galls on canes. Galls are reddish on red- and blue-fruited cultivars, green on white-fruited cultivars. Small holes in cane near gall. Cane is weakened and may drop off at the gall. Prune off and destroy infected canes a few inches below gall.

Grape phylloxera: These aphids infest roots and leaves and form galls. Galls on roots prevent movement of water and nutrients. Pea-size galls grow on the underside of the leaves. Western phylloxera do not make galls on leaves. Plant hybrid grapes that are resistant to this pest.

KIWIS

Leafroller caterpillars: Leaf edges rolled; caterpillars eat foliage, weakening the plant. Handpick if infestation is light. If heavy, spray with BTK when leafrollers are small.

Soft scales: Suck plant sap, weakening the plant. Scale honeydew attracts black sooty mold. If population is light, scrape from wood and leaves. Release predators such as the beetle *Chilocorus nigritus* or lady beetles. Spray with dormant oil in early spring before vines leaf out. Use rotenone or pyrethrin if other methods fail.

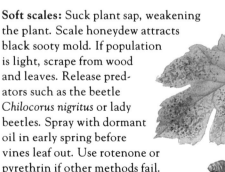

Citrus

ORANGES

Mites: Suck plant juices, weakening the tree. Leaves show silver stippling. Hose tree with plain water. Spray with insecticidal soap or release predatory mites. As a last resort, spray with superior oil or sulfur.
Also attack: Grapefruit, lemons, limes

Navel orange worms: Larvae bore into cracked fruit and spin cocoons. Pick off infested fruit and destroy. Pick and destroy dropped fruit. Spray BTK (*Bacillus thuringiensis* var. *kurstaki*) when larvae are hatching from eggs on fruit.
Also attack: Grapefruit, lemons, limes

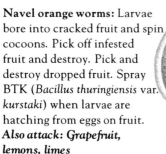

LIMES

Scales: Suck plant sap. Leaves turn yellow and drop; tree declines in vigor. Scales also excrete honeydew, which attracts sooty mold. Identify the scale species (it could be one of many) and release the appropriate predator. Spray with dormant oil before bud break. As a last resort, spray with superior oil, but do not apply after July. After July, use pyrethrin or rotenone instead.
Also attack: Grapefruit, lemons, oranges

LEMONS

Aphids: Suck sap, weakening the tree, and excrete honeydew that attracts black sooty mold. Grow small-flowered plants like dill or yarrow nearby to attract beneficials. Release lacewings. Place sticky bands on trunk to trap ants, which may be killing aphid predators. Spray dormant oil to kill overwintering eggs. Spray insecticidal soap or, as a last resort, pyrethrin or rotenone.
Also attack: Grapefruit, limes, oranges

GRAPEFRUIT

Mealybugs: Suck sap, weakening the tree. Leaves turn yellow and shrivel; fruit may drop early. Mealybugs excrete honeydew, which attracts sooty mold. Grow small-flowered plants like dill nearby to attract beneficials. Release mealybug destroyer (*Cryptolaemus montrouzieri*) or *Leptomastix dactylopii*, a parasitic wasp. Alternatively, spray with insecticidal soap.
Also attack: Lemons, limes, oranges

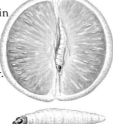

Preventing Diseases

Preventing plant diseases involves wise environmental management, good spacing, thorough sanitation and diligent pest control. It can also include using sprays to increase the plant's resistance to pathogens or disable the disease organisms.

Disease can be one of the most serious problems you will face with fruit plants. Unlike most pest insects and mites, many diseases are unobtrusive until they are doing real damage. Topical bacterial and fungal pathogens can often be treated and killed. However, viruses and pathogens that affect a plant's vascular system are sometimes impossible to eradicate. Prevention is the most effective line of defense and involves a complete program of good cultural care.

Preventive sprays work to keep plants from contracting diseases, not to cure infection. Consequently, you have to begin spraying in the early spring and keep it up, spraying at biweekly intervals through the season.

Compost tea: Compost tea, or watery compost extract as it is sometimes called, is gaining a great deal of support in scientific literature. Gardeners have used it for years, but it was not until the 1980s that researchers began proving that it protects plants against many fungal diseases. Organisms in the tea prey upon the pathogens or, in some cases, disrupt a pathogen's ability to chemically break down plant tissue.

To use compost tea as a disease suppressant, spray every two weeks through the season.

However, be aware that you are also adding nutrients, so it is possible to overfertilize plants this way. Stop spraying immediately if plant growth seems sappy or unnaturally dark.

BOTANICAL SPRAYS

Fermented nettle tea, liquid seaweed and equisetum (horsetail) tea also suppress fungal diseases if used before infection occurs. Fermented nettle tea contains high levels of trace elements and nitrogen, so take care not to overfertilize with this material. In addition

---

MINERAL CONTROLS

Minerals are not cure-alls. They do not affect viruses and control only a few bacteria. They are most effective against fungal diseases. They rarely kill the invading pathogen. Instead, they inhibit the germination of new fungal spores or limit the spread of the fungus or bacteria.

- **Bordeaux mix:** Controls downy mildew, peach leaf curl and black rot on grapes. Use as a dormant spray against anthracnose and fire blight.
- **Copper compounds:** Control anthracnose, apple scab, bacterial spot, cedar-apple rust, downy mildew and black rot on grapes.
- **Sulfur:** Prevents fungal spores from germinating. Use at the beginning of an infection and repeat to get control. Try to spray during cool periods. Controls apple scab, cherry leaf spot, powdery mildew and brown rot.
- **Lime sulfur:** Extremely caustic, so wear goggles and rubber gloves. Use only as a dormant spray to control apple scab, powdery mildew, peach leaf curl, cane blight and black knot.

to watching for leggy or unnaturally dark growth, pay attention to aphids, as an overfed plant is likely to be infested. Liquid seaweed contains high levels of trace elements but lower nitrogen levels than nettle tea. Horsetails do not contain such high nutrient levels as nettles (or seaweed), but they do contain a great deal of silica. Some growers believe that this mineral protects plants by strengthening their tissues. It is unlikely that excess nutrient levels will be created by spraying with horsetail tea on a biweekly schedule.

Equisetum tea: Mix ⅛ cup of dried horsetails, *Equisetum arvense*, in 1 gallon of unchlorinated water in a stainless steel pot. Bring this mixture to a boil and then turn it down to a slow simmer for half an hour. Then strain the liquid into glass jars. Store the jars in a cool, dark cupboard. To use, dilute the liquid to make a 5- to 10-percent solution.

Sprays from the Kitchen

Baking-soda sprays and garlic-oil sprays can kill some fungi on contact.

Baking-soda spray: Mix 1 teaspoon of baking soda in 1 quart of water. Add 1 teaspoon of insecticidal or mild liquid dishwashing soap like Ivory liquid. Use without diluting.

Garlic-oil spray: Mince 3 ounces of very fine garlic and steep in 2 teaspoonsful of mineral oil for 24 hours. Add 1 pint of water and ¼ ounce of liquid soap. Mix well and strain. Use by mixing 1 or 2 tablespoons of liquid with 1 pint of water.

Horticultural Oil Sprays

- **Leafy plants:** Use a 1 percent solution for disease control. Mix 2½ tablespoonsful of oil with 1 gallon of water.
- **Dormant plants:** Use a 2 percent solution to kill many overwintering pests. Mix 5 tablespoonsful of oil with 1 gallon of water.

Spray Safely

- **Protective clothing:** Wear protective clothing, rubber gloves and a face mask. Spray early on a cloudy day and only in calm weather to avoid drift. Thoroughly clean the sprayer after use. Use a duster made for mineral fungicides.
- **Spray thoroughly:** Spray both sides of leaves or, if using as a dormant spray, make certain that it penetrates cracks and crevices in the bark.

Commercial Sprays and Dips

Oil has been used to kill insects for many years, primarily as a spray applied when the plants were dormant. Today, a new class of oils has come onto the market that are so highly refined they don't damage the foliage of most plants. They are still used primarily as insecticides. However, researchers working with them noticed that they also killed downy mildew, either when used alone or in combination with baking soda. Try out the spray on only a few leaves at first. If the leaves show damage within three or four days, do not spray whole plants.

Highly refined horticultural oils are called "superior," "summer" or "supreme" oils. To control diseases on leafy plants, apply the oils only when there is a plentiful supply of soil moisture and when temperatures are above freezing but below 85°F. Do not apply horticultural oils if sulfur has been used on the plants in the last month. Do not apply sulfur for a month after using an oil either because the combination kills leaves.

Galltrol-A: This newly released material prevents roots from contracting crown gall. It contains a nonpathogenic bacterium (*Agrobacterium radiobacter*) that colonizes the roots without doing them any damage. Crown gall organisms cannot inhabit plant tissues that host this bacterium.

Common Fruit Diseases

In the fight against disease, correct identification is essential for successful control. If a plant has a fungus, for example, you will need to know whether to treat it with copper, baking soda or sulfur.

It is important to identify viral diseases as they cannot be eradicated. Most viral infections are transmitted by sucking insects such as aphids, so covering healthy plants with a row cover may prevent a virus from being moved around the garden. In the case of a debilitating disease, it is best to destroy infected plants.

TREES

APPLES

Apple scab: Brown to metallic black fungal spots, which become scabby on leaves and fruit. Fruit may be deformed; fruit and leaves may drop early. Choose resistant cultivars. Rake up leaves in fall. In a warm, wet spring, spray weekly with copper or sulfur, from bud break until 3–4 weeks after petal fall.

Cedar-apple rust: Leaves show pale yellow fungal spots that turn orange. Fruit is small and deformed and may drop early. Choose resistant cultivars. Avoid planting within 4 miles of junipers.

Fire blight: Bacterium causes leaves to blacken and growing tips to wither and bend down. Branches may show sunken cankers. Choose resistant cultivars. Prune off infected branches 6 inches below infection site. Don't overfertilize plants.
Also attacks: Sour cherries, pears

Summer disease: Rotting spots on fruit caused by several fungal pathogens. Choose resistant cultivars. During the dormant season, prune off spurs and branches showing cankers. In fall, collect and destroy all mummified fruit.

APRICOTS

Bacterial canker: Branches wilt and die; amber gum may ooze from bark. Lesions are apparent on bark. Choose resistant cultivars. Prune out infected branches 6 inches below lesions. Spray with copper or Bordeaux mix during late summer, after harvest or in early fall.
Also attacks: Sour cherries

Brown rot: Forms brown circular mold on fruit. Fruit rots inside and drops prematurely. Thin fruits so they don't touch. Remove some leaves to improve air circulation in the tree's canopy. Remove diseased and fallen fruit. Spray with sulfur if many fruits show symptoms.

TREES

CHERRIES, SOUR

Brown rot: Blossoms turn brown and rot; fruit develops brown spots that enlarge and become fuzzy in humid conditions. Remove mummified fruit in fall and winter. Prune off any gummy twigs and branches. Spray with copper or sulfur just before blooms open and again later in the season. Protect fruit from insect damage.
Also attacks: Sweet cherries

Cherry leaf spot: Leaves develop circular reddish to purple fungal spots that fall out, leaving holes. Tree declines in both productivity and winter hardiness and may die in time. Choose resistant cultivars. Rake up leaves in fall and spread compost under tree. Spray with sulfur in early spring and repeat 7–10 days later.
Also attacks: Sweet cherries

Powdery mildew: Leaves are covered with white powdery fungal growth. Look for this disease when warm days follow cool nights in midsummer. Spray with baking-soda solution (see page 88). Spray with sulfur weekly as necessary.

FIGS

Leaf blight: Causes fungal spots on leaves. Leaves and twigs turn yellow and die. Rake up and destroy fallen leaves and fruit. Prune out to remove infected twigs and increase air circulation.

Rust: Causes leaves to develop numerous yellow fungal spots and die. Rake up and destroy all fallen leaves and fruit. Spray weekly with sulfur or copper if rust symptoms appear.

NECTARINES

Brown rot: Fungal spots on flowers; fruit develops spots and mold; fruit loses interior texture before ripe. Twigs show sunken cankers, and leaves wilt, turn brown and die. Choose resistant cultivars. Spray compost tea every 10–14 days from early spring to harvest. Prune off twigs and branches with lesions in winter. Destroy shriveled, rotten fruit. Spray sulfur or lime-sulfur, as for peach scab (below).
Also attacks: Peaches, plums

Peach scab: Scabs on young fruit; malformed fruit; skin may crack. Rake up and destroy leaves, fruit and old mulch in fall. Spread new mulch. In spring, remove mulch, spread compost and reapply mulch. From the time flower buds show green until they open, spray sulfur every 10–21 days. Spray until harvest if wet or humid and fungal spots appear on fruit.
Also attacks: Peaches

TREES

PEACHES

Bacterial leaf spot: Sunken spots on fruit; small cracks may occur. Leaves develop yellowish green water-soaked spots, turn yellow and drop. Choose resistant cultivars. Spray sulfur or lime-sulfur at first symptoms and again 10 days later or after heavy rain.

Peach leaf curl: Leaves pucker, curl down and turn reddish, with powdery gray fungal spores; new growth is stunted and may die. Fruit drops or has rough skin. Choose resistant cultivars. Spray compost tea or liquid seaweed twice a week through season. Spray lime-sulfur after leaf drop and in spring when buds swell or at first symptoms, then again in 2–3 weeks.

PEARS

Fire blight: Flowers and leaves wilt and look scorched. Branches may develop sunken cankers and die. Ensure good air circulation by pruning and planting in an open position. Prune to 1 foot below infection site. Sterilize tools between cuts. Spray copper sulfate when dormant. Spray streptomycin or copper just before flowers open and every 4–7 days until fruit forms.
Also attacks: Quince

Pear scab: Weakens tree. Forms corky fungal scabs on leaves and fruit or causes malformed fruit. Choose resistant cultivars. In fall, rake up and destroy leaves, old mulch and dropped fruit. Prune out infected twigs in late winter. Spray lime-sulfur early in growing season.

PEARS, ASIAN

Fabraea leaf spot: Causes small, dark, round fungal spots with purple margins on the leaves. Leaves yellow and drop, weakening tree. Spray copper when leaves have broken from buds and are beginning to enlarge, then once or twice more at 2-week intervals.

Pseudomonas leaf blight: Bacterial disease causes leaves to turn very dark in fall. Tree is weakened. Choose resistant cultivars. Prune as for fire blight (above). Sterilize tools between cuts.
Also attacks: Quince

PLUMS

Black knot: Galls appear on branches. Tree may become stunted and die. Choose resistant cultivars. Prune off infected growth 6 inches below the tarlike fungal gall. Sterilize tools between cuts. Spray Bordeaux mix before bud break. Remove wild plums and cherries growing nearby, since they may be carriers.

PERSIMMONS

Anthracnose: Fruit and leaves covered with purplish, brown or yellow spots. Weakens tree and ruins fruit. Spray Bordeaux mix or lime-sulfur in later winter. Spray copper during outbreaks. Rake up and destroy infected leaves.

Fruit-Bearing Bushes

BLUEBERRIES

Mummyberry: Fungus causes berries to shrivel up. Pick off and destroy infected berries. Rake ground under bushes to remove fallen berries and leaves. Spread new mulch each fall and again in spring.

Stem canker: Oozing fungal sores weaken and finally kill stems. Choose resistant cultivars. Make sure soil drainage is very good. Top dress with good-quality, well-finished compost yearly in early spring.

CURRANTS, GOOSEBERRIES AND JOSTABERRIES

Anthracnose: Small fungal spots on foliage; leaves yellow and drop early. Spray compost tea from early spring through frost as a preventive measure. Spray copper if symptoms occur. Spray when dormant with Bordeaux mix. Rake up leaves and old mulch in fall and lay new mulch.

Leaf spot: Round or irregular yellow or brown fungal spots on leaves, or brown or black spots with yellow margins. Spray Bordeaux mix just after plants leaf out. If spots appear, spray copper every 10–14 days.

Powdery mildew: Powdery fungal patches on shoots and fruit. Prune to keep plant open. Spray compost tea from spring through frost as a preventive measure. Spray light infections with baking-soda solution (see page 88) or heavy infections weekly with sulfur.

ELDERBERRIES

Powdery mildew: Powdery fungal spots spread to cover leaves. Spray compost tea from early spring through frost to prevent infection. If infection occurs, spray superior horticultural oil or baking-soda solution (see page 88).

Stem and twig cankers: Dark fungal lesions become sunken and may ooze. Infected branches may weaken, break and eventually die. Prune out and destroy infected canes. Prune to improve air circulation.

STRAWBERRIES

Powdery mildew: Powdery fungal spots on leaves; fruit rots before ripening. Increase spacing by digging out excess runners. Spray compost tea every 2–3 weeks from spring until frost. At first sign, spray sulfur.

Red stele: Roots rot and appear red when sliced across. Plant strawberries where none have grown for 10 years. Check with your Extension agent to learn which regionally adapted cultivars are resistant to this fungus.

Verticillium wilt: Plants turn yellow, drop leaves and die. Do not plant where any member of the tomato family has grown for 3 years. There is no cure, so destroy fungus-infected plants. Ask your Extension agent about resistant cultivars.

BRAMBLES

BLACKBERRIES

Anthracnose: Fungus causes spots on foliage; weakens and may girdle canes. Prune for good air circulation. Spray compost tea every 7–14 days while in leaf. Spray lime-sulfur when dormant. Prune out infected canes. Rake up and burn leaves and fruit in fall.
Also attacks: All brambles

Crown gall: Bacterial disease causes corky swellings on roots, crown and canes before they die. Dig up and destroy infected plants. Choose resistant cultivars. Treat roots of new plants with Galltrol-A.
Also attacks: All brambles

Orange rust: Fungal pustules appear in spring on leaves and canes. Incurable. Dig up and destroy canes immediately. Do not confuse with a less-serious rust that may appear in fall.
Also attacks: All brambles

Verticillium wilt: Leaves turn yellow and die, weakening growth. Do not plant where members of the tomato family or strawberries have grown within 3 years. Choose resistant cultivars. Solarize soil before planting.
Also attacks: All brambles

Botrytis fruit rot: Fruit is covered with gray fungal mold, becomes water-soaked and rots. Prune for good air circulation. Pick and destroy infected fruit. Spray compost tea to prevent infection in subsequent seasons.
Also attacks: Dewberries, loganberries, raspberries, tayberries

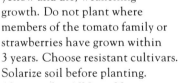

Powdery mildew: Powdery fungal spots spread to cover leaves and fruit. Prune for good air circulation. Spray superior oil midseason or spray sulfur or baking-soda solution weekly in midseason.
Also attacks: Loganberries, raspberries, tayberries

Crumbly berry virus: Fruit crumbles when ripe. Control aphids, which transmit viruses (see page 84). Incurable, so dig up and destroy infected plants.
Also attacks: Raspberries, tayberries

Mosaic virus: Chlorophyll (green pigment) is killed. Irregular yellowing occurs between veins. Control aphids, which transmit viruses (see page 84). Incurable. Dig up and destroy infected plants.
Also attacks: Loganberries, raspberries, tayberries

RASPBERRIES

Anthracnose: This fungus causes purplish spots on foliage. It weakens canes and flower stalks of black and purple raspberries. For control measures, see blackberries above.

Spur blight: Fungal spots near buds. Stunted laterals; leaves drop. Mostly affects red raspberries. Prune for good air circulation. Spray compost tea every 7–14 days when in leaf. Spray when dormant with lime-sulfur. Prune out infected canes; rake up and burn dropped foliage and leaves.
Also attacks: Tayberries

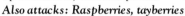

VINES

AMERICAN GRAPES

Botrytis fruit rot: Fruit becomes covered with a fuzzy-looking gray fungal mold. Prune to increase air circulation. Spray compost tea on vines and leaves every 7–10 days, beginning in spring. Destroy infected fruit. *Also attacks: European, hybrid and muscadine grapes*

Pierce's disease: Common in the South. Leaves and fruit become scorched and eventually plant dies. Get professional diagnosis if you think your vine is infected with this virus, since there is no cure and you will have to dig out the plant. Control grape leafhoppers, which transmit disease (see page 85). *Also attacks: European, hybrid and muscadine grapes*

Powdery mildew: Most common in the western U.S. Powdery fungal mold on leaf surfaces. Fruit ripens unevenly and may split. Choose resistant cultivars. Spray compost tea every 7–14 days when in leaf. Spray baking-soda solution or sulfur in early spring before flowering. *Also attacks: European, hybrid and muscadine grapes*

EUROPEAN AND HYBRID GRAPES

Anthracnose: Fruit and leaves show sunken fungal spots that are dark on the margins and light in the centers. Stem tips rot. Prune off and destroy all infected fruit and canes. Spray in early spring, before vines leaf out, with lime-sulfur. Many hybrids are resistant. *Also attacks: Muscadine grapes*

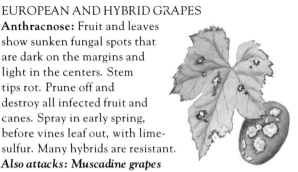

MUSCADINE GRAPES

Black rot: Common in the Southeast. Most severe in warm, humid weather. Fruit shows brown fungal spots before shriveling and drying. Leaves may wilt; canes may show dark lesions. Choose resistant cultivars. Keep area weed-free. Prune for good air circulation. Remove and destroy infected fruit and canes. Spray copper or Bordeaux mix early in spring if infection is serious. Respray once a week or after rain until midsummer. Do not apply copper while flowers are open.

KIWIS

Botrytis blight: Leaves and fruit develop water-soaked fungal spots, become covered with fuzzy gray or whitish mold and rot. Use preventive sprays of compost tea weekly from early spring until fall. Prune to eliminate excess growth, increasing air circulation. Prune off and destroy all infected leaves, flowers and fruit.

Crown rot: Slows plant growth; leaves turn yellow. Vine begins to rot at soil level. Plant kiwis only on well-drained sites. Do not mulch close to the base of the vine.

CITRUS

ORANGES

Brown rot gummosis: Dark spots ooze amber-colored gum. Lower branches may die. Plant trees grafted onto resistant rootstock. Do not let irrigation water splash onto trees. Scrape away cankers on bark to expose healthy tissue; allow to dry and then spray with copper.
Also attacks: Grapefruit, lemons, limes

Citrus scab: Fruit and leaves show light brown, corky areas. To prevent infection, plant only in well-drained soil in areas where air circulation is good. Do not let irrigation water splash on trees.
Also attacks: Grapefruit, lemons, limes

LEMONS

Greasy spot: Weakens the tree and pits fruit skins. Collect and destroy leaf litter. Remove mulch and replace with fresh material. Do not wet tree when irrigating. Spray with copper or superior oil if problem persists.
Also attacks: Grapefruit

Melanose: Leaves and fruit develop sunken spots that turn brown. Prune out dead wood that may be harboring fungal spores. Damage is primarily cosmetic. Spray with copper if disease seems to be taking over, but otherwise, don't worry.
Also attacks: Grapefruit, limes, oranges

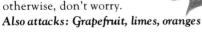

LIMES

Anthracnose: Leaves develop yellow, brown or purplish fungal spots. Small twigs and branches die. Fruits of some cultivars develop brown spots. Rake up and destroy infected leaves. Prune off diseased wood 6–12 inches below obviously infected area. Spray with Bordeaux mix or lime-sulfur while plants are dormant.
Also attacks: Grapefruit, lemons, oranges

GRAPEFRUIT

Brown rot: Fruit develops a white fungal mold or firm light brown spots on skin. This disease spreads from spores on ground. Prune off low-hanging branches. Remove and destroy infected fruit. Spray copper on low-hanging branches and on the soil surface under tree. Mulch around tree with water-absorbent mulch like straw, grass clippings or shredded leaves. Spray with sulfur before flowers open and again after petals drop.
Also attacks: Lemons, limes, oranges

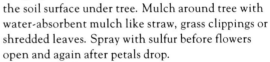

Animal Pests

Birds and squirrels beat you to your fruit. Deer come and prune uninvited. Bark is gnawed by rodents and shredded by cats. Fortunately there are ways to protect your fruit—and leave some for the animals, too.

Animals can do a great deal of damage to fruit crops. Gardeners vary in their approach to protecting their fruit. Some people net trees to exclude birds, while others try to harvest most of the fruit and leave a few for the birds. A few gardeners plant fruits especially for the birds. Mulberries and chokecherries are often used this way because birds prefer them to other berries.

RODENTS

Mice, rats and rabbits are the most destructive rodents in an orchard. In regions where heavy snows bury the seedheads of plants, protection is mandatory. Rodents love mulches, tunneling under them to hide from their many predators and also for warmth. Pulling mulches 6 inches back from the trunks and canes of fruit plants discourages rodents slightly because they don't want to cross an open area, where they are vulnerable to predators.

Plastic tree guards (see page 49), hardware-cloth "fences" and heavy tree-wrap tape all prevent rodents from eating tree bark. But if rodent pressure is likely to be heavy, a double protection of a tree wrap or guard plus a hardware-cloth fence is sensible. Some trees should be mulched above the graft union during winter to protect the union from the freeze-thaw cycle. In such cases, double protection is imperative.

CATS

Although cats do catch rodents, they can damage fruit plants. A young tree has just the right texture and tension for a scratching post, but too much scratching will shred the bark irreparably. Fortunately, tree guards protect trunks from overzealous claw sharpening.

Cats like kiwi vines almost as much as they like catnip, and it affects them in a similar way. Be prepared for a wholesale onslaught. Protect these vines with a screen of hardware cloth set close enough to keep the cat from jumping inside but far enough away so the cat can't reach the bark through the holes.

BIRDS

Birds' thieving ways with small fruit are legendary—and for good reason. Cherries, blueberries, blackberries, raspberries—you name it, they like it, and their harvest timing is perfect.

Nets are the most foolproof bird protectors. Garden-supply houses sell various sizes so you can choose what you need based on the size of your plants. Row-cover material like Reemay is also useful because it's easy to throw over a bush. Chicken wire works as long as birds can't reach the fruit through the mesh.

Scare-eye balloons frighten birds that are prey for owls as long as you keep moving them around your planting. They must be suspended at least 8 to 10 feet above the ground to convince the birds that the giant eyes really could be attached to a predator.

Bird Scares

▲ **Scare eyes** *are big, heavy-gauge plastic balloons with "eyes" painted on them. Birds think that predators are watching them.*

▲ **Aluminium pie plates** *don't add much to the visual appeal of a garden, but their glitter, movement and noise scare off birds.*

◄ **Christmas icicles** *can be recycled in the summer garden. Tie bunches together and hang on stakes above berries, or tie them to tree branches.*

Birds are also reluctant to come too close to sparkling, flickering light because it reminds them of fire. Aluminum pie plates, scare tape and aluminum foil all deter birds. Set up these devices so that they reflect sunlight when they move in the wind. Scare tape should be reeled out to form a spiral, left slack enough so light patterns continue to change and arranged to enclose the planting. Staple the ends to wooden stakes or tie it to supports. Aluminum foil strips can be wrinkled and lightly tied to the bushes. Mylar Christmas tree "icicles" can be tied in bunches and attached to stakes so they overhang a planting. Pie plates must swing in the breeze to be frightening.

Deer

Deer are the bane of many a commercial orchardist. They come in to eat fruit drops in the summer and generally return during the fall and winter to do a little pruning. Fences are the most effective way to deal with deer. However, not everyone can put up the sort of fence that keeps them out. They can jump most fences that seem appropriate for small yards and gardens. An electric fence about 6 feet tall is usually effective. But if you live in a town or city, installing an electric fence may be illegal. You are left with three alternatives: you can train your dog to patrol the area, you can build a fence around each fruit tree (see page 49) or you can use deer repellents.

Repellents: Commercial repellents, which repel deer by smell and/or taste, are fairly effective as long as you reapply them after heavy rains. You can make your own spray by blending two raw eggs in a cup of water, then mixing the solution with gallons of water in the sprayer. This spray loses its potency and also washes off in rain, so you must remember to replenish it every few weeks, especially during the winter months.

Soap is one of the most effective deterrents. Buy a case of the smelliest deodorant soap available (see "Traps, Tricks and Barriers" on page 75). Without removing the wrapper, bore a hole through one end, string wire through the hole and suspend the soap from a branch. Use six to eight bars per small tree and even more for a full-size tree.

It is important to begin a deer deterrent program as soon as you plant. You want to keep deer away from the start. Since you can't hang soap from a whip, you are better off putting a deer cage around newly planted trees until they grow some branches.

Chapter

4

FRUIT DIRECTORY

~~~~

NOW THAT YOU *are familiar with all the basic techniques of growing fruits and berries successfully, you are in the exciting position of choosing what you want to grow. The "Fruit Directory" contains over 30 of the most popular fruits for home gardens. The recommended cultivars have been carefully selected so you'll find fruits that are suitable for every zone.*

For easy reference, the fruits are arranged in alphabetical order. Each entry is packed with specific information you need to grow your chosen fruits to perfection, using only organic and ecologically sound methods.

HOW TO USE THE FRUIT DIRECTORY

The sample layout, displayed below, explains the symbols used, as well as all the essential points that are covered for every fruit.

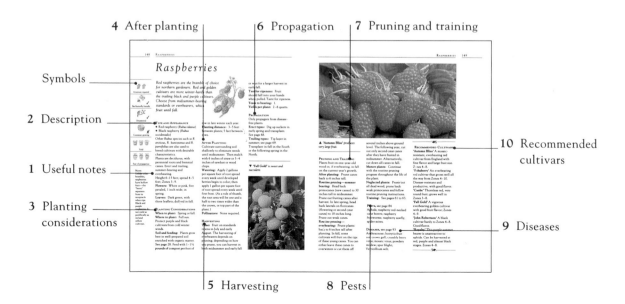

4 After planting   6 Propagation   7 Pruning and training

Symbols

2 Description

1 Useful notes

3 Planting considerations

10 Recommended cultivars

9 Diseases

5 Harvesting   8 Pests

UNDERSTANDING THE SYMBOLS

Attention required
Is the plant easy or demanding to grow?

Easy, requiring little attention

Easy but requires regular attention

Moderate

Demanding

Bee and butterfly friendly
Does the plant attract bees and butterflies?

  ✓ ✗
Yes   No

Ornamental value
Is the plant ornamental as well as fruit bearing?

 ✓ ✗
Yes   No

Suitability for container growing
Can the plant grow and bear fruit successfully in a container?

 ✓ ✗
Yes   No

Yield
Does the plant fruit heavily, moderately or lightly?

Light

Moderate

Heavy

Propagation
Is the plant easy or difficult to propagate?

Difficult

Moderate

Easy

1 Useful notes
Useful information, not otherwise covered in the entry, is provided in margin notes.

2 Description
• Different species available.

• The figures given for the heights and spreads are the minimum and the maximum for the particular fruit. In the case of many tree and bush fruits, this figure will be governed by the choice of rootstock and/or cultivar. The number of years a plant takes to reach its maximum size also varies, from a few months in the case of a strawberry to 20 years or more in the case of a standard tree grafted on a vigorous rootstock.

• The zone rating given for each fruit and cultivar refers to the zone range on the United States Department of Agriculture Plant Hardiness Zone Map, reproduced on page 154.

3 Planting considerations
The advice given on when and where to plant, soil preparation, feeding and planting distances will help you choose the best fruit or cultivar for your yard and give your plants the best possible start. Conditions that aren't ideal may not result in complete failure, but your plants will not yield as well and are more likely to suffer from pest and disease problems.

4 After planting
To keep your plants in the best possible health and bearing abundantly, follow the advice given on routine cultivation, mulching, watering and pollinators.

5 Harvesting
To harvest fruits at their peak, testing for ripeness is all-important. Years to bearing are given for every fruit to give you an idea of how long you will have to wait to pick that first juicy orange or pear. Yields-per-plant guidance will help you plant enough for your family. Tips on storage are also given where appropriate.

6 Propagation
The easiest methods of propagation are given for every fruit.

7 Pruning and training
Detailed notes are provided for pruning after planting, routine pruning and pruning mature plants. There are tips on bringing neglected plants back into bearing. Finally, alternative training methods are listed, with a cross-reference to the details of the appropriate technique.

8 Pests
The most common foes are listed alphabetically, with a guide to where to find symptoms, controls and remedies.

9 Diseases
The most common diseases — fungal, viral or bacterial — are listed alphabetically, with a guide to where to find symptoms, controls and remedies.

10 Recommended cultivars
There is a selection of some of the best cultivars along with descriptions.

Apples

Attention required

 ✓

Bee/butterfly friendly

 ✓

Ornamental

 ✓

Container growing

Yield

Ease of propagation

NOTE
Apples require from 1,200 to 1,500 chilling hours. Check with your nursery or supplier to learn the chilling requirements for the cultivars you are choosing.

Bite into a homegrown apple. The taste is so much better than those bought at the store that you'll wonder what took you so long to grow your own. If you have space, you can plant cultivars for dessert, cooking and winter storage, and enjoy a range of apple flavors.

TYPE AND APPEARANCE
• Deciduous tree (*Malus* spp.) Height 6–30 feet; spread 8–40 feet; Zones 3–9.
Flowers: White and pink, star-shaped, 1 inch wide, generally in clusters; in early spring.
Leaves: Glossy green, pointed and serrated; turn dull yellow or brown in fall.

PLANTING CONSIDERATIONS
When to plant: Early spring or fall.
Where to plant: Full sun, preferably on a slope. Don't plant where apples have grown before.
Soil and feeding: Well-drained, deep, moderately rich soil with a pH of 6.5–6.8. See page 28. In late winter, apply 5–10 pounds of compost around each tree.
Planting distance: Spacing depends on the plant's rootstock, but generally space plants at least as far apart as the height of a mature tree—25–30 feet for a standard tree, 12–15 feet for semidwarf and 6–8 feet for a dwarf.

Rootstocks: Affect plant height, spread and vigor. See page 24.

AFTER PLANTING
If you're planting a dwarf apple, support the tree with a durable metal or wooden stake. Keep a 4 by 4-foot area around the tree clear of weeds. Mulch with 4–6 inches of straw or 3–4 inches of a denser material such as hardwood bark chips. Leave at least 6 inches of bare earth between the trunk and mulch material to protect against rodent damage. To protect against winter sunscald and/or rodent damage, wrap the tree with a white plastic tree guard or paper tape. Trunks can also be painted white with diluted latex paint to protect from sunscald and protected with a wire mesh or hardware cloth screen "sleeve" to deter rodents.
Watering: Apply approximately 2 gallons per week per square foot of root area. (As a rule of thumb, the root area will be one and a half

to two times wider than the crown, or top part of the plant.)
Pollinators: Check with your nursery or supplier to make sure you buy pollinators that are compatible and that bloom at the same time as the plants they're to fertilize.

HARVESTING
Time: Summer cultivars ripen in late July through August. Storage types ripen in late August through late October and can take light frost.
Test for ripeness: Color should be well developed. Taste summer cultivars to test for ripeness. Gather storage apples when size and color indicate ripeness. Pick with stems for best storage quality. Some winter types, such as 'Idared', do not develop full flavor until a month or so after harvesting.
Years to bearing: 3–8, depending on rootstock and cultivar.
Yields per plant: 10–30 bushels, depending on age and size of tree.

PROPAGATION
Graft in early spring, bud in summer. See pages 70 and 71.

PRUNING AND TRAINING

Plants fruit on one- to ten-year-old wood, on spurs or tips of branches depending on the cultivar.

After planting: Prune according to final shape desired—central leader, modified central leader, open or espalier.

If pruning to a single stem (a whip) for a central leader or a modified central leader, remove all side branches and prune back the top to about 30 inches.

If pruning to an open center, an option for standard trees, prune back the central leader (the future trunk) to about 24 inches and, if possible, select the first

scaffold branches. Trim this to two buds, cutting above an outward-facing bud. Remove all other branches.

Whatever shape you've chosen, rub or pinch off unwanted new shoots during the season.

At one year: For a central leader or modified central leader form, select and retain three or four well-spaced scaffold branches with wide crotch angles, about 6–8 inches apart on the trunk. Ideally, the scaffold branches will form a "spiral staircase." Prune off all other branches and cut back the leading branch to about 1 foot above the desired location of the next set of scaffold branches.

▲ **'Freedom'** *is disease-resistant.*

Always train branches into wide angles by weighting them with rocks or clothes pins or, for a very small branch, bracing a round-end toothpick between the trunk and branch.

For a vase-shaped or open-center form, select and retain three or four scaffold branches, as described, and trim them back by a third, cutting above an outward-facing bud. The lowest scaffold branch should be about 15 inches from the ground. If there are fewer than four good scaffold branches, leave the central leader in place to produce more scaffolds. However, if you have four strong, well-placed and wide-angled branches, trim back the central leader at least 3 feet from the ground. Always train branches into wide angles as described above.

At two years: For central leader or modified central leader forms, follow directions for "At One Year," selecting additional scaffold branches. Prune off unwanted branches. Cut back scaffolds to a third of their length, always above an outward-facing bud.

For vase-shaped or open-center trees, if necessary, select the remaining branch(es) to make up the four scaffolds and trim back the central leader to about 3 feet above the ground. If you did this the previous year, prune off the weakest side branches from the scaffolds. Trim back all branches to about a third of their length, always cutting above an outward-facing bud.

At three years: For the central leader form, select two or three additional scaffold branches and trim back to a third of their length. Select and retain two or three side branches on the previous year's scaffold branches, including one near the end of the branch, and prune off all others. Trim back these side branches by about a third, to an outward-facing bud. Prune off all branches that grow downward or into the center of the tree.

For a modified central leader form, cut back the central leader if the tree now has an adequate number of scaffold branches (six to eight for a standard tree, four to six for a dwarf or semidwarf). If there aren't enough scaffold branches, let the leader continue to grow and cut it back the following season. Prune side branches as described for the central leader form.

For vase-shaped or open-center trees, select and retain the strongest and best-placed two or three branches on each scaffold, including a shoot near the tip of the scaffold. Remove the remaining shoots.

Routine pruning: For all forms, remove damaged and malformed fruits and thin excess fruits, leaving them 6 inches apart or one fruit per spur. Remove water sprouts and suckers. Maintain desired shape by removing excess growth, always cutting out

◀ '**Empire**' *is a good choice for long-term storage.*

▲ **'Granny Smith'** *is a favorite for fresh eating.*

branches that grow downward or into the center of the tree.
Mature plants: Remove dead, broken, crossing or competing branches.

Neglected plants: Take at least two years to renovate these trees. In the first year, cut out dead, weak and poorly positioned branches, suckers and water sprouts. In the second year, prune to shape and trim back overly long growth.

Training: Plants can also be trained as a cordon, fan or espalier. See pages 58 to 59.

PESTS, see page 80
Apple maggot fly; codling moth; plum curculio.

DISEASES, see page 89
Apple scab; cedar-apple rust; fire blight; summer disease.

RECOMMENDED CULTIVARS

Cultivars vary, so check with your local Extension agent and suppliers to learn which ones are best for your climate. Also ask about appropriate pollinators.

'Empire' Red fruit. Stores until February in home conditions. Hardy in Zones 5–9 and in protected spots in Zone 4. Resistant to fire blight and cedar-apple rust, but easily contracts scab.

'Freedom' Hardy in the Northeast and quite disease resistant. Medium-size, red fruit for fresh eating as well as for making sauce. Zones 3–8.

'Granny Smith' This crisp and juicy dessert apple requires warm climates and long seasons. Zones 6–9.

'Liberty' Disease-resistant with dark red-striped skin. Stores well until January in home conditions. Excellent flavor. Easy to prune and train. Zones 4–8.

'Redfree' A summer cultivar with shiny red fruit, hardy in protected areas in Zone 4. Immune to scab and cedar-apple rust; resistant to fire blight. Stores for one to two months. Zones 4–8.

'Sweet 16' Large, red fruit with yellow undercoat and excellent flavor. Resists scab but contracts fire blight and cedar-apple rust. A late bloomer, it is a good choice for windy, cold Zone 4 conditions. Tends to grow upright and must be carefully trained. Zones 3–8.

'William's Pride' A disease-resistant summer cultivar. Fruit is dark red, very juicy and has a tart undertone to its sweetness. Does not keep long into the winter. Zones 5–7.

Apricots

Attention required

Bee/butterfly friendly ✓

Ornamental ✓

Container growing ✓

Yield

Ease of propagation

NOTE
Apricots require between 700 and 1,000 chilling hours. Check with your nursery or supplier for each cultivar's chilling requirements. In cold climates with fluctuating temperatures, dwarf trees can be grown in containers and moved to protected spots if frost threatens the blossoms or fruit.

Who wouldn't want to experience the sweet subtle flavor of apricots fresh from a tree — a flavor that's so rare in purchased fruit? Growing your own will enable you to make a treat such as apricot nectar. In sunny, protected spots in Zone 3, some apricots will fruit.

TYPE AND APPEARANCE
- European apricot (*Prunus armeniaca*)
- Manchurian apricot (*P. armeniaca* var. *mandschurica*)
- Siberian apricot (*P. armeniaca* var. *sibirica*)

Deciduous tree.
Height 4–25 feet; spread 6–25 feet; Zones: 3–8 (Manchurian and Siberian), 5–9 (European).
Flowers: White and/or pink, fragrant, 1 inch wide, five-petaled; in very early spring.
Leaves: Deep green with reddish undertones when young; oval, 1–2 inches long; turn bright yellow in fall.

PLANTING CONSIDERATIONS
When to plant: Spring.
Where to plant: Full sun, in a spot protected from wind and early frosts such as a north-facing slope.
Soil and feeding: Well-drained, sandy loam soils with moderate to high fertility. Apricots are pH-tolerant. See page 28. In late winter, spread 5–10 pounds of compost under each tree.

Planting distance: Equal to height of mature tree, roughly 25 feet apart for standard trees and 12–15 feet apart for dwarfs.
Rootstocks: Affect height, spread and vigor. See page 24.

AFTER PLANTING
Cultivate to keep weed-free, and mulch with 6 inches of straw or 3–4 inches of a denser material such as wood chips.
Watering: Apply at least 2 gallons per week per square foot of root area. (As a rule of thumb, the root area will be one and a half to two times wider than the crown, or top part of the plant.)
Pollinators: Many European apricots are self-fruiting, but bear more heavily with a pollinator. Those of Manchurian and Siberian origin require pollinators. Nanking cherry works well as a pollinator.

HARVESTING
Time: July and August.
Test for ripeness: Harvest when hanging fruit has softened somewhat but before it drops to the ground. Once harvested, apricots won't continue to ripen. Fruit ripens over two or three weeks. Handle carefully, since apricots bruise easily.
Years to bearing: 3–5.
Yields per plant: 1–2 bushels (dwarf cultivars); 3–4 bushels (standard-size European cultivars).

PROPAGATION
Graft in early spring; bud in summer. See pages 70 and 71.

▼ *'Moorpark' bears large fruit.*

◄ **'Moongold'** *is a relatively new and very hardy cultivar. Choose it along with 'Sungold' for a pollinator if you live in the North.*

RECOMMENDED CULTIVARS
'Blenheim' A European cultivar. With 'Royal', this is the most popular commercial cultivar because of its reliable annual fruiting, good flavor and appearance. Zones 7–10 for 'Blenheim' and 'Royal'.
'Harglow' A relatively late-blooming European cultivar that grows only 18 feet tall. Excellent flavor and very juicy. Zones 7–10.
'Harlayne' A European cultivar with relatively good disease resistance. Smaller fruit than many cultivars, but excellent flavor. Zones 7–10.
'Moongold' A Manchurian cultivar with excellent cold-hardiness, high yields and fine flavor. Zones 4–10.
'Moorpark' A European cultivar grown in the Northeast. The large fruits ripen over a longer period than most apricots. Zones 5–9.
'Sungold' Another excellent Manchurian cultivar, generally planted with 'Moongold' for cross-pollination. Zones 4–10. Although hardy, 'Sungold' and 'Moongold' are more reliable in the Midwest, with its steady cold, than in areas in the East, with fluctuating temperatures.

PRUNING AND TRAINING
Plants fruit on one-year-old spurs and branch tips.
After planting: Prune to a modified central leader or open-center form. See page 101 in the Apples entry.
Routine pruning: Thin developing fruit to at least 2 inches apart if fruit is very heavy. Prune as for a modified central leader or open-center form; see pages 55 and 102.
Mature plants: To stimulate new growth, remove any branches that are more than three years old.
Neglected plants: Cut out old branches and those that are poorly positioned or weak.
Training: Plants can be trained as a cordon, fan or espalier. See pages 58 and 59.

PESTS, see page 80.
Codling moth; plum cucurlio.

DISEASES, see page 89.
Bacterial canker; brown rot.

Blackberries

*These juicy berries are easy to grow.
Not long ago, it was difficult to grow
blackberries outside Zones 6–8,
but today you can find cultivars that
flourish from Zones 5–10. Thornless
cultivars are usually trailing, while most
prickly types are upright.*

Attention required

Bee/butterfly friendly ✓

Ornamental ✓

Container growing ✗

Yield

Ease of propagation

NOTE
Blackberries
are good
companions
to grapes
because they
host
beneficials
that control
grape
leafhopper
and two-
spotted spider
mite.

TYPE AND APPEARANCE
● Standard blackberry (*Rubus*
spp.)
● Thornless blackberry
(*Rubus ulmifolius*)
Plants are deciduous, with
perennial roots and erect or
trailing biennial canes.
Height 5–8 feet, if pruned and
trained like a raspberry; 15 feet
for trailing cultivars; spread
4–5 feet, if pruned and
trained; Zones 5–10,
depending on the cultivar.
Flowers: White with pink
undertones or soft pink, five
petaled, 1 inch wide; in
spring.
Leaves: Dark green, with
three leaflets; dusky red in fall.

PLANTING CONSIDERATIONS
When to plant: Spring or fall.
Where to plant: Full sun; in
northern areas, protect from
cold winds.
Soil and feeding: Plants grow
best in well-prepared soil
enriched with organic matter.
See page 28. Feed with 1–1½
pounds of compost per foot of
row in late winter each year.

Planting distance: 3–5 feet
between plants; 6 feet between
rows.

AFTER PLANTING
Cultivate surrounding soil
shallowly to eliminate weeds
until midsummer. Mulch with
6 inches of straw or weed-free
hay in midsummer.

Watering: 2 gallons per
square foot of root spread,
applied weekly through
midsummer. (As a rule of
thumb, the root area will be
one and a half to two times
wider than the crown, or top
part of the plant.) Decrease to
1 gallon per week while
developed berries are
ripening. To protect against
fungal diseases, do not wet
leaves or fruit.
Pollinators: None required.

▼ **Flowers** *on this 'Chester'
blackberry promise dense clusters
of large fruit.*

HARVESTING

Time: Late summer.

Test for ripeness: After berries are black, gently pull fruit to see if it comes off the cane easily; taste for sweetness. Harvest every other day throughout the season.

Years to bearing: 1.

Yields per plant: 1–2 quarts.

PROPAGATION

Tip layer in summer (see page 69.)

PRUNING AND TRAINING

Plants bear fruit on one-year-old wood.

After planting: Head back canes to 6 inches. As the canes grow in summer, tie them to a wire trellis.

Routine pruning: Head primocanes (new canes) back to 30 inches in midsummer. Prune out fruiting canes after harvest. In late spring, head back laterals on floricanes (flowering or second-year canes) to 18 inches long. Prune out weak canes and thin remaining canes to stand 8–10 inches apart.

Mature plants: Continue with the routine pruning program throughout the life of the plant.

Neglected plants: Prune out all dead wood, prune weak primocanes, head back primocanes in midsummer (as described above) and follow routine pruning instructions.

Training: See pages 61 to 63.

▶ **'Darrow'** *produces amazing quantities of fruit.*

PESTS, see page 84

Aphids; raspberry red-necked cane borers; two-spotted and European red spider mites.

DISEASES, see page 93

Anthracnose; crown gall; orange rust; Verticillium wilt.

RECOMMENDED CULTIVARS

'Chester' Thornless; hardy to Zone 5; high yields.

'Darrow' Produces large fruit; hardy to Zone 5.

'Dirksen' Resists many fungal diseases; Zones 5–10.

'Lawton' A good choice for Zones 8–10; resists crown gall.

'Loch Ness' Thornless; hardy to Zone 5; tolerates high humidity.

'Navajo' Thornless; good choice for Zones 6–9.

Blueberries

Perfect for eating out of hand, making into preserves and pies, canning and freezing for the winter, blueberries are as versatile as they are delicious. Five plants will provide enough berries for most families. If you want more, take some cuttings. Blueberries are simple to propagate.

Attention required

Bee/butterfly friendly ✓

Ornamental ✓

Container growing ✓

Yield

Ease of propagation

NOTE
Highbush and lowbush cultivars require 700–1,000 chilling hours each winter, while rabbiteyes need only 100–500, depending on the cultivar.

TYPE AND APPEARANCE
- Highbush blueberry (*Vaccinium corymbosum*) Height 5–6 feet; spread 5–6 feet; Zones 4–8.
- "Half-high" blueberry (*Vaccinium corymbosum* cultivars) Height 2–4 feet; spread 2–4 feet; Zones 4–7.
- Lowbush blueberry (*Vaccinium angustifolium*) Height 1–3 feet; spread 5–8 feet; Zones 3–8.
- Rabbiteye blueberry (*Vaccinium ashei*) Height 15–18 feet; spread 5–6 feet; Zones 6–9.
Deciduous shrubs.
Flowers: Pinkish white, ¼-inch-long bells, in clusters; in early spring.
Leaves: Dark glossy green, smooth, pointed tip, small (¾–2½ inches long); turn red in fall.

PLANTING CONSIDERATIONS
When to plant: Spring or fall.
Where to plant: Full sun for highbush and half-high cultivars. In hot climates or with lowbush cultivars, afternoon shade is desirable. Rabbiteyes are the best choice for warm climates.
Soil and feeding: Highbush and lowbush blueberries need a permanently moist, very acid soil (pH 4.0–5.0) that drains freely and is high in organic matter. Rabbiteyes will grow in drier soil with a pH of 5.5. Soil acidity can be increased before planting by digging in sulfur (see page 18) and well-rotted organic matter, such as peat moss, composted sawdust, pine needles or oak leaves. Where drainage is poor, plant blueberries in a raised bed or in containers. Feed blueberries in spring by applying 1 pound of bloodmeal on top of the mulch for every 10 feet of row.
Planting distance: Highbush and half-high cultivars: 5–6 feet apart in row; 8 feet between rows.
 Lowbush: 5–6 feet apart in row; 8 feet between rows.
 Rabbiteye: 8 feet apart in row; 8–10 feet between rows.

AFTER PLANTING
Mulch with 4 inches of acidic organic mulch such as rotted pine needles or partially rotted leaves from hardwood trees. Remove and compost mulch each fall, replacing with a fresh layer.
Watering: Apply 2 gallons of water per square foot of root area weekly. (As a rule of thumb, the root area will be one and a half to two times wider than the crown, or top part of the plant.)
Pollinators: Plant at least three different cultivars to ensure good pollination.

HARVESTING
Time: Late summer to fall. Leave berries on bushes for about a week after they turn blue so sugars will develop.
Test for ripeness: Berries should drop easily into your hand. Taste for sweetness and ease of picking daily.
Years to bearing: 4–5.
Yields per bush: 3–8 quarts.

PROPAGATION
Take hardwood cuttings when plant is dormant and root in nursery bed in early spring; see page 71.

PRUNING
Plants fruit on two-year-old wood.
After planting: Prune off any damaged wood. Remove all flower buds to give the bush time to establish itself.

Routine pruning: In early spring, prune out winter-damaged or diseased wood and burn it well away from blueberry bushes. Once the plant is dormant, remove laterals that produced berries.

Rabbiteye bushes require less pruning than highbush and half-high cultivars, so just prune out damaged or diseased wood, as described above.

At four to five years: Remove the oldest, weakest branches from highbush and half-high cultivars. To increase fruit size, remove the tips of the remaining branches to leave three to five buds per branch. Do not prune off tips of rabbiteye cultivars because this is where berries form.

Lowbush cultivars need severe thinning to remain productive. Once the plant is four years old, you can begin to thin the many laterals to increase the yield of large, healthy berries.

Mature plants: For highbush and half-high cultivars, cut out approximately one in three of the old branches to stimulate new growth. Remove branches that prevent light and air from reaching the center of the plant before cutting others away.

If the berries of lowbush cultivars become small and yields decline, cut off all stems about ½ inch above the soil surface while they are dormant in early spring, and let them regrow through the season.

Neglected plants: Cut all branches to within a few inches of the ground. In subsequent years, prune as described above for mature plants.

PESTS, see page 83
Birds; blueberry maggot; cherry fruitworm.

DISEASES, see page 92
Mummyberry; stem canker.

RECOMMENDED CULTIVARS
'Cabot' A very old and very productive lowbush cultivar. Plant with 'Greenfield', another old and productive cultivar, for pollination.
'Early Bluejay' Vigorous plant with good fruit quality and yields. Often produces berries in third year.
'Herbert' Late-ripening with high yields, very dark-colored, large sweet fruit that are resistant to cracking.
'Meader' Excellent fruit quality and size. Hardy to −25°F with snow cover. Fruit resistant to cracking and dropping.
'Northblue' Hybrid cross with highbush and lowbush parentage. Hardy to −30°F (Zone 3) with snow cover. High yields. Grows only 2 feet tall.
'Southland' Rabbiteye cultivar. Compact bush with late-ripening, flavorful fruit.
'Tifblue' Rabbiteye cultivar, good companion to 'Southland'. Vigorous, erect bush with very good fruit quality.
'Wolcott' Appropriate for mid-Atlantic states and southeastern United States. Vigorous plant with high yields of fine-flavored fruit.

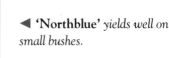

◀ **'Northblue'** *yields well on small bushes.*

Boysenberries

The large, dark fruits of boysenberries taste like wild blackberries — not surprising when you know that boysenberries have blackberry, loganberry and raspberry in their parentage. They are ideal for sandy soils, being more drought-tolerant than most brambles.

Attention required

  ✓

Bee/butterfly friendly

  ✓

Ornamental

  ✗

Container growing

Yield

Ease of propagation

TYPE AND APPEARANCE

● Standard, thornless boysenberry (*Rubus ursinus* var. *loganobaccus* cv.) Plants are deciduous, with biennial canes and perennial roots.
Height 6–10 feet, depending on trellising design; spread 3–5 feet; Zones 6–8.
Flowers: White with pink undertones or soft pink, five petaled, 1 inch wide in spring.
Leaves: Dark green, with three leaflets; dusky red in fall.

▲ **Boysenberries** *grow on vigorous trailing plants.*

PLANTING CONSIDERATIONS

When to plant: Spring or fall.
Where to plant: Full sun; in cool areas, protect from winter winds.
Soil and feeding: Same as for blackberries (see page 106). Boysenberries grow well in sandy soil.
Planting distance: 8 feet between plants; 8 feet between rows.

AFTER PLANTING

Cultivate surrounding soil shallowly to eliminate weeds until midsummer. Mulch with 6 inches of straw or weed-free hay in midsummer.
Watering: 2 gallons per square foot of root spread, applied weekly through midsummer. (As a rule of thumb, the root area will be one and a half to two times wider than the crown, or top part of the plant.) Decrease to 1 gallon per week while developed berries are ripening. To protect against fungal disease, do not wet leaves or fruit.
Pollinators: None required.

HARVESTING

Time: Late summer and fall.
Test for ripeness: After berries are fully colored, gently pull a fruit to see if it comes off the cane easily; taste for sweetness. Harvest every other day through the six- to eight-week season.
Years to bearing: 1.
Yields per plant: 4–8 quarts.

PROPAGATION

Tip layer (see page 69).

PRUNING AND TRAINING

Same as for blackberries (see pages 61 to 63).

PESTS, see page 84

Aphids; raspberry red-necked cane borers; two-spotted and European red spider mites.

DISEASES, see page 93

Anthracnose; crown gall; orange rust; Verticillium wilt.

RECOMMENDED CULTIVARS

'Giant Thornless' Slightly less vigorous and less tolerant of cold conditions than thorny types; Zones 5–10.
'Old-Fashioned' Vigorous, thorny and high yielding, with large fruit; Zones 5–9.
   Buy from a nursery in your region to make sure you get a strain that's well adapted to your climate.

Cherries, Sour

Sour cherries make the best pies, preserves and sauces. And some "sour" cultivars are sweet enough when ripe to eat fresh—just ask birds or children. Sour cherries tolerate both colder and hotter climates and grow into smaller trees than sweet cherries. They're also easier to train.

TYPE AND APPEARANCE
- Sour cherry (*Prunus cerasus*) Deciduous tree.
Height 20 feet (standard), 8–10 feet (genetic dwarf); spread 25 feet (standard), 10–12 feet (dwarf); Zones 4–8.
Flowers: White, ¾–1 inch wide, five-petaled, in clusters; in early spring.
Leaves: Dark green, oval, toothed, 3–5 inches long; turn yellow in fall.

PLANTING CONSIDERATIONS
When to plant: Spring; fall in mild-winter areas.
Where to plant: Full sun. Choose a protected spot in Zones 4 and 5 to prevent frost damage on blooms and young fruit. A north-facing slope is ideal.
Soil and feeding: Soil should be well drained, with moderate fertility and a pH of 6.0–6.8. See page 28. Apply 5–10 pounds of compost around each tree in late winter.
Planting distance: 25 feet (standard); 15 feet (dwarf).

Rootstocks: Affect height, spread and vigor. See page 24.

AFTER PLANTING
Keep area clear of weeds. Apply 6 inches of straw mulch or 3–4 inches of denser material like wood chips.
Watering: Apply 2 gallons per week per square foot of root area. (As a rule of thumb, the root area will be one and a half to two times wider than the crown, or top part of the plant.)
Pollinators: None required.

HARVESTING
Time: Mid-July to early August.
Test for ripeness: Taste for ripeness when fruit has colored.
Years to bearing: 4–5.
Yields per plant: 2 bushels (standard); 1 bushel (genetic dwarf).

PROPAGATION
Graft in early spring; bud in summer. See pages 70 and 71.

PRUNING AND TRAINING
Plants fruit on one- to ten-year-old spurs.
After planting: Prune to an open-center form or modified central-leader form. See page 101 in the Apples entry. If pruning to an open-center form, cut back leader 25–30 inches above ground level.
Routine pruning: Establish a strong framework of well-positioned branches. Thin side branches only when necessary to keep air circulation high. Train tree to form wide crotches, as branches tend to grow vertically. See page 54.

▼ **Sour cherries** *ready for picking.*

Attention required

Bee/butterfly friendly

Ornamental

Container growing

Yield

Ease of propagation

Mature plants: Remove weak, dead or poorly positioned branches. Cut back branches if growing too high or long. Do not thin fruit unless set is extremely heavy.

Neglected plants: Thin out old, weak and dead branches. Cut back branches to make tree manageable.

Training: Plants can also be trained as fans or espaliers; see pages 58 and 59.

PESTS, see pages 80 and 81
Birds; black cherry aphids; cherry fruit fly maggot; pear slug (pear sawfly); plum cucurlio.

DISEASES, see pages 89 and 90
Bacterial canker; brown rot; cherry leaf spot; powdery mildew.

▶ **'Montmorency'** *makes memorable pies.*

RECOMMENDED CULTIVARS

'English Morello' A small tree with branches that may need propping or heading back. Good for the North. Dark red fruit is tart, late and crack-resistant. Zones 4–8.

'Evans' A new, very hardy cultivar with large red-fleshed fruit. Good for the Midwest and Northeast. Zones 4–8.

'Meteor' A genetic dwarf with large fruit that has red skin and yellow flesh. Resistant to leaf spot. Good for the Northeast, since it is unusually hardy. Zones 4–8.

'Montmorency' The standard by which sour cherry fruits are judged. Large, crack-resistant, red-skinned fruit with firm yellow flesh. Blooms quite early. Good for Zones 6–9.

'Northstar' A genetic dwarf with red flesh that resists cracking, cherry leaf spot and brown rot. Early ripening. Particularly good for the Midwest. Zones 4–8.

Cherries, Sweet

Wouldn't you love to pick ripe sweet cherries from a tree in your own yard? These luscious fruits can be eaten fresh or preserved. In spring, cherry blossoms are breathtakingly lovely, and the shiny bark looks good any time of year.

TYPE AND APPEARANCE
● Sweet cherry (*Prunus avium*)
Deciduous tree.
Height 25–35 feet (standard), 10–15 feet (semidwarf), 6–8 feet (dwarf); spread 35–40 feet (standard), 18–20 feet (semidwarf), 8–15 feet (dwarf); Zones 5–9.
Flowers: White, in large showy clusters; in early spring.
Leaves: Dark green, toothed, 3–6 inches long; turn yellow in fall.

PLANTING CONSIDERATIONS
When to plant: Spring; fall in mild-winter areas.
Where to plant: Full sun, in very well-drained soil. (A north-facing slope is ideal.)
Soil and feeding: Soil should be well drained and deep, with good fertility and a pH of 6.0–6.8. See page 28. Apply 5–10 pounds of compost around each tree in late winter.
Planting distance: At least 30–35 feet (standard); 20 feet (semidwarf); 15 feet (dwarf).

Rootstocks: Affect height, spread and vigor. See page 24.

AFTER PLANTING
Keep root area weed-free. Mulch with 6 inches of straw or 3–4 inches of denser material like wood chips, making sure to leave 6–8 inches of bare soil between trunk and mulch.
Watering: Apply 2 gallons of water per week per square foot of root area. (As a rule of thumb, the root area will be one and a half to two times wider than the crown, or top part of the plant.) Do not overwater when fruit is ripening, as many sweet cherry cultivars crack easily.
Pollinators: Most sweet cherries require pollinators; check with your nursery or supplier for appropriate matches, since not all sweet cherries are compatible.

HARVESTING
Time: Early to late July.
Test for ripeness: Taste for ripeness when fruit has colored.

Years to bearing: 4–5.
Yields per tree: 2 bushels for standards; 1 bushel for semidwarfs; ¾ bushel for dwarfs.

PROPAGATION
Graft in very early spring; bud in summer. See pages 70 and 71.

PRUNING AND TRAINING
Plants fruit on one- to ten-year-old spurs. (Spurs remain fruitful for up to 10 years.)
After planting: Begin pruning to open-center form. See page 101 in the Apple entry. Cut leader back to 35 inches above ground level. Cherries must be trained to spread (see page 138); their natural habit is upright.
Routine pruning: Prune for balanced growth that is well spaced around the trunk. Do not let scaffold branches form directly opposite each other because their weight may cause the trunk to split. After main framework branches have developed, cherries need little pruning. Do not thin fruit or flowers.
Mature plants: Prune out dead and weak branches. Remove crossing branches, and cut back growth that threatens to imbalance the pruning pattern.
Neglected plants: Cut back excessively long branches, prune out dead, weak and poorly positioned branches.

Attention required

✓ Bee/butterfly friendly

✓ Ornamental

✓ Container growing

Yield

Ease of propagation

NOTE
Sweet cherries require at least 1,000 chilling hours. They do not perform well where summer temperatures regularly exceed 90°F.

▲ **'Stella'** *is a favorite and is self-pollinating.*

Training: Plants can also be trained as espaliers. See pages 58 and 59.

PESTS, see page 80
Birds; cherry fruit fly maggot; pear slug (pear sawfly).

DISEASES, see page 90
Brown rot; cherry leaf spot.

RECOMMENDED CULTIVARS

Recent sweet cherry cultivars are more resistant to cracking from heavy rains at ripening.
'Bing' The cultivar everyone thinks of with sweet cherries. Good flavor, but highly susceptible to cracking if it rains heavily while fruit is ripening. Zones 5–9.
'Black Tartarian' A very early sweet cultivar with dark, medium-size fruit of good flavor. Good pollinator for 'Bing'. Zones 5–9.
Duke cherries (*P.* × *effusus*) Crosses between sweet and sour cherries, the fruits of these cultivars look like sweet cherries and taste like sour cherries. The trees are as hardy and as small as most sour cherries. Most cultivars have 'Duke' somewhere in their name. They need a late-blooming sweet cherry pollinator. Zones 4–8.
'Rainier' A recent yellow sweet cherry cultivar with flavor that many people prefer to 'Royal Ann' and the other older yellows. Zones 5–9.
'Sam' A dark red sweet cherry that resists cracking. Bears early and heavily with flavorful fruit. Zones 5–8.
'Stella' A self-fruitful sweet cherry cultivar that can also pollinate most other cherries. Zones 5–9.

Citrus

Imagine picking oranges and other fresh citrus fruits from trees outside your own backdoor! These handsome trees enhance any yard with their glossy leaves, brightly colored fruits and fragrant white flowers.

TYPE AND APPEARANCE

- Grapefruit (*Citrus* × *paradisi*)
Height 25–35 feet (standard), 10–15 feet (dwarf); spread 20–35 feet (standard), 10–15 feet (dwarf); Zones 9–10.
- Lemon (*Citrus limon*)
Height 10–25 feet (standard), 5–10 feet (dwarf); spread 10–20 feet (standard), 5–10 feet (dwarf); Zones 9–10.
- Lime (*Citrus aurantiifolia*)
Height 15–20 feet (standard), 7–10 feet (dwarf); spread 15–25 feet (standard), 8–12 feet (dwarf); Zone 10.
- Sweet oranges, blood oranges, navel oranges (*Citrus sinensis*)
Height 20–25 feet (standard), 8–10 feet (dwarf); spread 20 feet (standard), 10 feet (dwarf); Zones 9–10.
Evergreen trees.
Flowers: White, waxy, extremely fragrant, 1 inch wide; borne throughout the year.
Leaves: Dark, glossy green, 3–5 inches long.

PLANTING CONSIDERATIONS

When to plant: Early spring.
Where to plant: Plant oranges and grapefruit in full sun. Lemons and limes prefer full sun, but will tolerate some shade in hot areas. Do not plant any citrus in a lawn.
Soil and feeding: Citrus need soil with high fertility that is moisture retentive but well drained. They need a pH of 6.0–6.5 and can't tolerate salty soil. Citrus will grow well in a raised bed. See page 17. Feed citrus by spreading compost or well-rotted manure around each tree, or scattering a few handfuls of bloodmeal or cottonseed meal under each tree. Feed trees every six weeks to four months beginning in February and ending in September or October.
Planting distance: Oranges — 25 feet (standard); 10–15 feet (dwarf); lemons — 20–30 feet; limes — 15–20 feet; grapefruit — 35 feet.

AFTER PLANTING

Keep root area weed-free. Mulch lemons, limes and grapefruits with newspapers covered with 2 inches of grass clippings, 2–3 inches of grass clippings alone or 3–4 inches of wood chips. Mulch oranges with 4 inches of grass clippings, sawdust, or wood chips or gravel in a circle extending at least 3 feet around each plant.
Watering: Apply 2–3 gallons of water per week per square foot of root area. (As a rule of

---

CONTAINER GROWING

Some lemon, lime and orange cultivars are naturally dwarf and adapt well to container culture; a 5-gallon pot is sufficient for some trees. During the summer, put the trees outdoors in a sheltered, partially shaded spot. Keep trees thoroughly watered. In the winter, move them indoors to a place where they will get at least half a day of sun. Mist leaves frequently to keep the humidity high, and maintain warm days (70°–75°F) and cool nights (45°–55°F). Apply liquid fertilizer such as seaweed extract at least once a month. Hand-pollinate with an artist's brush.

Attention required

Bee/butterfly friendly

Ornamental

Container growing

Yield

Ease of propagation

NOTE
Harvest all ripe fruit if frost threatens. When ripe fruit has been slightly frosted, it may still be usable if harvested immediately.
Protect trees from frost by wrapping them in old blankets; uncover when the weather warms.

From *Growing Fruits and Vegetables Organically*, (Rodale Press, 1994)

thumb, the root area will be one and a half to two times wider than the crown, or top part of the plant.) Keep the soil consistently moist.

Pollinators: None required.

HARVESTING

Time: Grapefruit — late fall and early spring; lemons — heaviest in spring and fall, with lighter harvests throughout the year; limes — heaviest in early summer, with lighter harvests throughout the year; navel oranges — early winter; sweet oranges — anytime; blood oranges — spring. Location may also influence harvest time.

Test for ripeness: Pick oranges, limes and grapefruit when fruit is full size and the skin is fully colored. Taste to test oranges for ripeness if you're unsure of the correct skin color. Most lemons are harvested when full size but still green, then allowed to cure in a place with good air circulation for 30 to 60 days. You can let them ripen fully on the tree, although they will have less juice and thicker skins. Remove fruit with pruners at stem attachment, taking care not to bruise skins. Ripe fruit of many cultivars will hang on the tree for several weeks after ripening without degrading.

Years to bearing: 2–4.

Yields per tree: Grapefruit — 6–8 bushels (standard), 2–3 bushels (dwarf); lemons and limes — 3–5 bushels (standard), ½–1½ bushels (dwarf); oranges — 4–6 bushels (standard), 1½–2 bushels (dwarf).

PROPAGATION

Graft in early spring, bud in summer. See pages 70 and 71.

PRUNING AND TRAINING

Plants fruit on the current year's wood.

After planting: Citrus are self-shaping and only need maintenance pruning. Some lime cultivars form shrubs rather than single-trunked trees. Prune these only as necessary to keep them healthy and to allow light into the center of the plants.

Routine pruning: Remove dead, weak or poorly positioned branches after harvest. Remove suckers from oranges through the year.

Mature plants: Simply continue with routine pruning. Keep center open to light and air. To stimulate new growth, remove old wood that no longer bears well.

Neglected plants: Prune off dead wood and older branches that no longer bear well. Cut back overly long branches. Do not prune too much during the first year.

Training: Plants can also be trained as fans, cordons or espaliers. See pages 78 and 79.

◀ **'Marrs'** *is a juicy orange for the backyard.*

Pests, see page 86
Aphids; mealybugs; mites;
navel orange worms; scales.

Diseases, see page 95
Anthracnose; brown rot;
brown rot gummosis; citrus
scab; greasy spot; melanose.

▶ **'Eureka'** *is a small and
thornless lemon tree that is
perfect for home gardens.*

Recommended Cultivars

It is important to buy plants locally, since rootstocks are chosen to suit local conditions.

Grapefruit
'Duncan' Grown in Florida, Louisiana and Texas. White-fleshed cultivar that is able to tolerate high humidity.
'Marsh' Most common white-fleshed grapefruit. Seedless. Grown in Arizona, California, Florida and Hawaii. Heat-tolerant.
'Red Seedless' Grown in Florida. Ripens in winter. Red-fleshed, seedless fruit.
'Ruby' Red-fleshed cultivar grown in California, Florida and Texas.
'Thompson Pink' Grown in Florida. Light pink rather than red fruit. Almost seedless.
Limes
Seedling limes are available from some nurseries. But since the fruit quality of seedlings is variable, choose a named cultivar rather than a seedling tree.

'Bear's Seedless' A juicy, seedless lime. Very thorny, relatively cold-hardy. Grown in Arizona and California.
'Mexican' Grown in California, Florida and Texas, this is the "Key lime" of the famous pie. Small yellow fruit.
'Persian' ('Tahiti') Grown in Florida and Hawaii. Good flavor; not cold-tolerant.
Lemons
'Eureka' Grown in Arizona, California and Louisiana. This is the lemon most often found in grocery stores. Small plant with few thorns. Prune to open-center form.
'Lisbon' Grown in Arizona, California and Florida. Heat-tolerant. Very thorny, but attractively shaped tree.
'Meyer' The most cold-tolerant cultivar, hardy to −17°F. Grown in Arizona, California, Florida, Louisiana and Texas. Excellent choice for containers.
'Ponderosa' A mild-flavored lemon, with huge fruits on a dwarf plant. Grown in

California. Another good container choice.
'Villa Franca' Grown in Arizona, Hawaii and Florida. Relatively large tree and fruit. Quite thorny.
Oranges
'Marrs' A sweet orange grown in Arizona. Fruit ripens in fall. Used for both juice and fresh eating.
'Robertson' Navel orange grown in California and Arizona. Ripens in winter and spring.
'Ruby' A blood orange grown in California, Florida, Louisiana and Texas. Very sweet and good color.
'Temple' Grown in Florida and some parts of Louisiana. A hybrid between a sweet orange and a mandarin. Winter-bearing.
'Valencia' The most common sweet orange in both California and Florida. Fruit ripens in summer and holds on the tree for long periods. A good juicer.

Currants, Gooseberries and Jostaberries

NOTE
Black currants are known to be alternate hosts for white pine blister rust; don't plant them if there are pines in your area. Some states still forbid you to grow any of these shrubs. To be on the safe side, check your state's restrictions first.

All these species are ornamental shrubs with fruit that can be eaten fresh or preserved. Jostaberries, a hybrid cross between currants and gooseberries, have nearly black fruit that is larger than gooseberries and just as flavorful.

TYPE AND APPEARANCE
- Black currants (*Ribes nigrum*); red and white currants (*Ribes sativum, R. rubrum* and *R. petraeum*)
- American gooseberry (*Ribes hirtellum*)
- European gooseberry (*Ribes uva-crispa*)
- Jostaberry (*Ribes nidigrolaria*)

Deciduous shrubs.
Height 3–5 feet; spread 3–5 feet; Zones 3–6 (currants), 3–7 (gooseberries and jostaberries).
Flowers: Currants have small greenish yellow to violet-red flowers, in grapelike clusters, with five inconspicuous petals; in spring. Gooseberries and jostaberries have flowers held in smaller, drooping clusters, with more distinct petals.
Leaves: Lobed, medium green, decorative, arranged alternately along the stem, sometimes clustered; turn yellow in fall.

PLANTING CONSIDERATIONS
When to plant: Fall.
Where to plant: Full sun in the North; in Zones 6–7, plant where bushes receive filtered afternoon shade.
Soil and feeding: Plant in well-prepared, slightly acid soil enriched with plenty of organic matter. See page 28. Spread strawy manure or 3 pounds of soybean or cottonseed meal beneath each plant in late fall to late winter each year.
Planting distance: 4–6 feet between plants; 6–8 feet between rows.

AFTER PLANTING
Mulch with 6 inches of straw or 4 inches of rotted sawdust, finished compost or other compact organic mulch.
Watering: Apply 2 gallons per week per square foot of root spread. (The root area will be one and a half to two times wider than the crown, or top part of the plant.)

Pollinators: Some black currants do require a pollinator. Gooseberries, jostaberries, and red and white currants do not require a pollinator, although fruit set is enhanced when more than one cultivar is grown.

HARVESTING
Time: Midsummer.
Test for ripeness: Pick currants slightly underripe for jellies, near fully ripe for jams and fully ripe for eating fresh and drying. Taste test the bottom berry on the cluster and if it is at the correct stage of sweetness, cut the whole cluster.

Pick gooseberries singly, watching for thorns. Follow ripeness guidelines for currants. Some people find green gooseberries too tart for eating fresh and use them only in pies and preserves, saving the pink ones for fresh eating.

For jostaberries, follow ripeness guidelines for currants. Pick singly when at the correct stage of ripeness.
Years to bearing: 2–3.
Yields per bush: Currants 3–5 quarts, gooseberries 4–6 quarts, jostaberries 4–6 quarts.

PROPAGATION
Take hardwood cuttings of one-year-old stems, 8–10

inches long, when the plant is dormant, and plant the cuttings so only two buds show above the soil surface (see page 71).

PRUNING

Plants fruit on one-, two- and three-year-old wood. Currants can be trained as cordons; see pages 60 and 61.

After planting: Cut tops back to 10 inches and prune out any damaged or crossing wood.

Routine pruning: Prune out weak shoots, leaving six to eight strong stems, in early spring when plant is dormant. Remember to prune to open up the center of the plant, letting air in to protect against diseases.

Mature plants: In early spring, cut out canes that have borne fruit for three years. Allow new canes to grow each year as replacements. Maintain plants so that only 10 to 16 stems are growing at any one time, keeping a balance between new wood and first-, second- and third-year wood. Keep bushes open to ensure free circulation of air at all times.

Neglected plants: Cut out all old wood. Select six strong young stems and begin pruning as for mature plants after three years.

PESTS, see page 83
Currant borers; currant fruit fly.

DISEASES, see page 92
Anthracnose; leaf spot; powdery mildew.

RECOMMENDED CULTIVARS

CURRANTS
'Red Lake' Red currant. High yields, fine-flavored, large berries. Usually produces in second year after planting.
'Wilder' Red currant. Large fruit; high yields.
'Prince Consort' Black currant. Very hardy; good yields; sweet, dark fruit.

GOOSEBERRIES
'Pixwell' Pink gooseberries. Large, easy-to-pick berries hang down several inches from canes, nearly thornless.

'Welcome' Green gooseberries make excellent pies and preserves. Ripened to pink, they are sweet enough for some to eat them fresh. Early ripening.
'Leepared' and **'Poorman'** Gooseberries. Resistant to powdery mildew and leaf spot.

JOSTABERRIES
Hybrids Thornless canes. Self-pollinating. Very high yields, hardy. Very pest- and disease-free.

▲ **'Red Lake'** *yields early and prolifically.*

Dewberries

The dewberry, with its intense flavor, is a favorite with gardeners. Untrained, the canes trail along the ground. Use as a groundcover if you wish, but trellising decreases incidence of diseases and pests and makes harvesting easier.

Attention required

Bee/butterfly friendly ✓

Ornamental ✓

Container growing ✗

Yield

Ease of propagation

TYPE AND APPEARANCE
● Trailing dewberry (*Rubus caesius*)
Plants are deciduous, with perennial roots and slender, trailing biennial canes. Height 6–8 feet; spread 4–6 feet; Zones 5–9.
Flowers: White with pink tones, five-petaled, ¾ inch wide; in spring.
Leaves: Lighter green than most brambles, with three leaflets; red and yellow in fall.

PLANTING CONSIDERATIONS
When to plant: Spring or fall.
Where to plant: Full sun.
Soil and feeding: Plants grow best in well-prepared soil enriched with organic matter. See page 28. Feed with 1–1½ pounds of compost per foot of row in late winter each year.
Planting distance: 5–6 feet between plants; 8 feet between rows.

AFTER PLANTING
Cultivate surrounding soil shallowly to eliminate weeds until midsummer. Mulch with 6 inches of straw or weed-free hay or 3 inches of rotted wood chips.
Watering: 2 gallons per square foot of root spread, applied every week until berries begin to color; then 1 gallon per square foot of root spread until frost. (The root area will be one and a half to two times wider than the top part of the plant.)
Pollinators: None required usually, but see 'Flordagrand' below.

HARVESTING
Time: July and early August.
Test for ripeness: After a gentle pull, fruit should "fall" into your hands. Taste for ripeness.
Years to bearing: 1.
Yields per plant: 2–4 quarts.

PROPAGATION
Tip layer in summer (see page 69).

PRUNING AND TRAINING
Same as for blackberries (see pages 61 to 63).

PESTS, see page 84
Aphids; raspberry red-necked cane borers; Japanese beetles; raspberry fruitworms; two-spotted and European red spider mites.

DISEASES, see page 93
Anthracnose; botrytis fruit rot; crown gall; orange rust; powdery mildew; Verticillium wilt.

▲ **Dewberries** *have a strong flavor — mostly sweet but with tart undertones.*

RECOMMENDED CULTIVARS
'Carolina' Resists leaf spot diseases. Zones 7–9.
'Flordagrand' Requires a pollinator such as 'Oklawaha'. Zones 7–9.
'Lucretia' This relatively old cultivar bears large fruit — up to 1½–2 inches long — that ripens early in July. Zones 5–9.

Elderberries

Not only do elderberries look lovely in a hedge or along a fence line but they also fruit abundantly. When you buy elderberry plants, be sure to ask whether the fruit is edible, since many cultivars are available that are purely ornamental.

TYPE AND APPEARANCE
● American or sweet elder (*Sambucus canadensis*) Deciduous shrub.
Height 6–12 feet; spread 5–6 feet; Zones 2–9.
Flowers: Small, white and fragrant, borne in showy clusters; in May and June.
Leaves: Bright green, toothed, 5–6 inches long, shaped like a broad lance.

PLANTING CONSIDERATIONS
Where to plant: Full sun. In southern areas, filtered shade in afternoon is necessary.
Soil and feeding: Plant in deep, well-prepared soils enriched with plenty of organic matter. See page 28. Apply a topdressing of compost each season before you mulch.
Planting distance: 5–6 feet between plants; 10–15 feet between rows.

AFTER PLANTING
Mulch with 6 inches of straw.
Watering: Apply 2 gallons per week per square foot of root spread.

Pollinators: Plant two cultivars for good pollination.

HARVESTING
Time: From early August through mid-September.
Test for ripeness: Wait until fruit is dark before tasting for sweetness. Cut whole cluster.
Years to bearing: 2–4.
Yields per bush: 12–15 pounds.

PROPAGATION
Take dormant hardwood cuttings, 8–10 inches long, in spring or take softwood stem cuttings; see page 71.

PRUNING
Plants fruit on tips of one-year-old shoots and two- and three-year-old wood.
After planting: Head back to strong bud.
Routine pruning: Prune out canes that are four years old, leaving five to nine canes per plant. Remove weak and diseased wood.
Mature plants: Same as for routine pruning.
Neglected plants: Cut out all old wood and then begin routine pruning.

PESTS, see page 83
Birds; elder shoot borer.

DISEASES, see page 92
Powdery mildew; stem and twig cankers.

RECOMMENDED CULTIVARS
'Adams 2' A more productive and slightly later cultivar than the old-fashioned 'Adams 1'. Good disease resistance. Zones 3–8.
'Johns' Vigorous, with flavorful berries. A pollinator for 'Adams'. Zones 3–8.

Attention required

Bee/butterfly friendly

Ornamental

Container growing

Yield

Ease of propagation

▼ **'Johns'** *makes delicious pies and jellies.*

Figs

Fig Newton cookies are most people's introduction to this delicacy. But you can grow your own outside, in large containers, for the summer; once fall arrives, wheel the container into a cool, bright room in the house. Or bring them into a cold, dark shed until spring.

Attention required

Bee/butterfly friendly ✗

Ornamental ✓

Container growing ✓

Yield

Ease of propagation

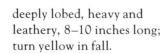

TYPE AND APPEARANCE
● Deciduous tree or large shrub (*Ficus carica*) Height 10–30 feet, depending on the climate and cultivar; spread 6–30 feet, also depending on the climate and cultivar; Zones 8–10 without winter protection, plus Zones 6–7 with winter protection.
Flowers: Borne inside the "fruit."
Leaves: Medium green, large, deeply lobed, heavy and leathery, 8–10 inches long; turn yellow in fall.

PLANTING CONSIDERATIONS
When to plant: Spring.
Where to plant: In full sun. In the North, plant against a white or stone south-facing wall.
Soil and feeding: Plant in well-drained soil. Figs tolerate a wide range of soil types and pH. See page 28. Spray figs with liquid seaweed monthly while plants are actively growing. Apply 5 pounds of compost per plant in late winter.

Planting distance: Depending on the climate and cultivar, plant 5–25 feet apart, so leaves of adjacent trees don't touch.

AFTER PLANTING
Mulch with moisture-retentive but nutrient-poor material like straw, since too much nitrogen discourages fruit formation.
Watering: Keep soil moist but not wet. Once fruits are enlarging, water every two or three weeks, applying 2 gallons of water per square foot of root area. (As a rule of thumb, the root area will be one and a half to two times wider than the crown, or top part of the plant.)
Pollinators: None are required for home garden cultivars.

HARVESTING
Time: June, for overwintered fruit; August for fruit formed the current year.
Test for ripeness: Pick when soft and fully colored.
Years to bearing: 3–6.
Yields per tree: 1 bushel (standard); 2–6 quarts (dwarf containerized).

WINTER PROTECTION
In the warmer half of Zone 7, protect figs in winter by wrapping plants with old blankets or burlap. In borderline climates, in Zones 6 and 7 as far north as Maryland, overwinter plants by burying them. In late fall, prune plants to 6 feet tall and tie the branches together to make a pillar. Dig a trench 2 feet deep and 6 feet long, starting beside the root ball. Line the trench with boards. Dig to expose the root ball on the side away from the trench. Wrap the plant in plastic and bend it into the trench. Fill the trench with straw or dried leaves, cover it with a board, then pile several inches of soil on top. If the tree is too rigid to bend, encase it with chicken wire, fill the wire cage with dry leaves and wrap the cage, including the top, with 6-mil plastic to shed snow and rain. Uncover plants when spring temperatures will go no lower than 30°F.

▶ **'Brown Turkey'** *is decorative as well as tasty.*

PROPAGATION
Propagate from suckers throughout the growing season. Take hardwood cuttings during the dormant season. See page 71.

PRUNING AND TRAINING
Plants fruit on tips of new growth and from leaf axils on one- and two-year-old branches.

After planting: Formative pruning differs with the type of fig grown. Most figs don't need formal pruning. Consult your local nursery or supplier for details.

Routine pruning: Check with your nursery or supplier to determine correct pruning procedure. With most cultivars, thin branches every two years during the dormant season. Do not prune back tips of 'Black Mission' and 'Celeste' because that is where the fruit forms. Cut back 'Texas Everbearing' to keep it in bounds.

Training: Plants can also be trained as espaliers. See pages 58 and 59.

PESTS, see page 81
Ants; spider mites.

DISEASES, see page 90
Leaf blight; rust.

RECOMMENDED CULTIVARS
Buy figs locally to get suitable cultivars for your region.

'Adriatic' Grown in California for drying. Green skin with pink pulp.

'Black Mission' Grown in California and the eastern part of the Deep South. Disease-resistant black fruit with red pulp.

'Brown Turkey' Grown in the East and used in containers. Fruit is brown with pink-colored pulp.

'Celeste' A somewhat hardy cultivar for the Southeast and for growing in containers. Bronze skin with amber-colored pulp.

'San Piero' A 'Brown Turkey' offspring grown in California with large fruits that have brown to purplish skin and pink pulp. Must be pruned back to remain fruitful and manageable.

'Texas Everbearing' Brown to purple-black fruit with pink pulp. Grown in the Gulf states and the Southeast. Resprouts if branches are killed in winter.

Grapes, American

Often referred to as bunch or fox grapes, American grapes can be grown in the eastern, midwestern and northwestern United States and in parts of southern Canada. Clusters are tight and skins slip off easily. Depending on the cultivar, they can be red, white or blue-black.

Attention required

Bee/butterfly friendly ✗

Ornamental ✓

Container growing ✗

Yield

Ease of propagation

NOTE
Pick grapes when slightly underripe for jellies. Pectin content declines as fruit sweetens, making the mixture harder to jell.

TYPE AND APPEARANCE
● American grapes (*Vitis labrusca*)
Perennial vine.
Height 12–15 feet; spread 3–5 feet, depending on pruning and training system; Zones 5–9.
Flowers: Small, in inconspicuous clusters; in spring.
Leaves: Dull to dark green, large, lobed; turn yellow in fall.

PLANTING CONSIDERATIONS
When to plant: Early spring or fall.
Where to plant: In full sun in an open site, ideally a south-facing slope, where air can circulate freely and drainage is good.
Soil and feeding: Plant in deep, well-drained, well-prepared soil with abundant organic matter incorporated. See page 28. Feed with 1–1½ pounds of compost per foot of row in late winter each year.
Planting distance: 8 feet between plants; 8–9 feet between rows.

AFTER PLANTING
Cultivate to remove weeds through the early part of the season. Mulch after midseason with 6 inches of spoiled hay or straw or 4 inches of a denser material, such as wood chips or cocoa bean hulls.
Watering: Water very generously to encourage deep rooting. From spring until the developed grapes begin to color, apply 4 gallons per square foot of root area every week to ten days. (As a rule of thumb, the root area will be one and a half to two times wider than the crown, or top part of the plant.) Then, after the grapes begin to ripen and until the end of the season, water only in serious drought conditions.
Pollinators: None required.

HARVESTING
Time: From mid-August through the fall, depending on the cultivar.
Test for ripeness: Pick the bottom grape on a bunch to taste for sweetness. Do not pick the bunch unless it tastes

sweet, since grapes do not ripen off the vine. Cut clusters with a stem "handle" still attached.
Years to bearing: 2–3.
Yields per plant: 10–15 pounds.

PROPAGATION
Take cuttings in early spring while the plant is still dormant. (See page 71 for hardwood cuttings.) Cuttings should be ½ inch in diameter and 18 inches long. Keep cool and dark until early spring. Plant in prepared ground, leaving only the top bud above the soil surface.

PRUNING AND TRAINING
Plants fruit on one-year-old wood.
After planting: Prune according to the four-arm Kniffen system (see page 64) or start training for an arbor.
Routine pruning: Continue to prune and train for the four-arm Kniffen system or on an arbor. American grapes do not fruit on buds near the main stem, so when pruning, leave 10–12 buds on each fruiting cane.
Mature plants: Prune and train for the four-arm Kniffen system or on an arbor.
Neglected plants: Cut out old wood, leaving only the trunk and four arms required for training to the four-arm Kniffen system. Follow

directions for that system in subsequent years.

PESTS, see page 85
Birds; grape berry moths (and larvae); grape leafhoppers; grape mealybugs; Japanese beetles.

DISEASES, see page 94
Botrytis fruit rot; Pierce's disease; powdery mildew.

▼ **'Canadice'** *shows good disease resistance.*

RECOMMENDED CULTIVARS

'Canadice' Very hardy red cultivar with good flavor and large clusters. Ripens in mid-August through most of Zones 4–8. Resistant to powdery mildew and downy mildew.
'Concord' The best-known American grape; a dark blue cultivar, hardy from Zones 4–9. Ripens by the end of September. A seedless selection is now available. Resistant to anthracnose, downy mildew, and botrytis fruit rot.
'Mars' A hardy blue cultivar with good disease resistance and tolerance of cold conditions. Ripens in mid-August to late August. Grow in Zones 5–8.
'Niagara' The best-known white cultivar, suitable for Zones 5–8. Big fruit from which the skin slips easily, making it a good choice for jelly and juice. Resistant to anthracnose and botrytis fruit rot.
'Reliance Seedless' A hardy red cultivar with good disease resistance. Ripens in early September to mid-September. Grow in Zones 4–8.

Grapes, European and Hybrid

Most cultivars grow well in California and the Southwest, but hybrids with mixed European and American ancestry are more vigorous than pure Europeans and, thanks to their greater resistance to the grape phylloxera aphid, helped save French vineyards during the late 1800s.

Attention required

Bee/butterfly friendly ✗

Ornamental ✓

Container growing ✗

Yield

Ease of propagation

NOTE
A grape leaf in a jar of dill pickles keeps them crisp during processing. Check Greek cookbooks for recipes for stuffed grape leaves. To preserve the leaves, put them in brine and refrigerate or freeze.

TYPE AND APPEARANCE
• European grape (*Vitis vinifera*)
Perennial vine.
Height 12–20 feet; spread 4–5 feet; Zones 7–9.
Flowers: Small, inconspicuous, borne in clusters; in spring.
Leaves: Dark to medium green, generally with rounded lobes, 6–8 inches long; turn yellow in fall.

PLANTING CONSIDERATIONS
When to plant: Early spring or fall.
Where to plant: In full sun in an open, well-drained site (ideally a south-facing slope).
Soil and feeding: Deep, well-prepared soil with abundant organic matter incorporated. See page 28. Feed with 1–1½ pounds of compost per foot of row in late winter each year.
Planting distance: 6–8 feet between vines; 10 feet between rows.

AFTER PLANTING
Cultivate the ground around vines to keep it clear of weeds, and don't mulch plants for the first year. Mulch with 6 inches of straw or 3–4 inches of a more compact material, such as wood chips or cocoa bean hulls, in midseason of the second year.
Watering: To encourage deep rooting, about 4 gallons per square foot of root area is required every week to ten days during the early part of the season. (As a rule of thumb, the root area will be one and a half to two times wider than the crown, or top part of the plant.) Do not water while grapes are coloring or after harvest unless there is a severe drought.
Pollinators: None required.

▶ **'Olivette Blanche'** *is a wine grape.*

HARVESTING
Time: Late summer to fall.
Test for ripeness: Taste bottom grape of cluster for sweetness. Do not harvest prematurely because grapes will not ripen off the vine.
Years to bearing: 2–3.
Yields per plant: 10–15 pounds.

PROPAGATION
Take dormant cuttings as described for American grapes; see page 124.

PRUNING AND TRAINING
Plants fruit on one-year-old wood.

After planting: Prune and train according to the four-arm Kniffen system on page 64 or train to a single wire.

Routine pruning: Prune during dormancy, cutting off most of the previous season's growth and leaving only two or three buds on each spur.

Mature plants: Maintain plant form by pruning out old "arms" as they lose vitality.

Neglected plants: Prune to the four-arm Kniffen system and maintain by following directions for that system in subsequent years.

Pests, see page 85
Birds; grape berry moths (and larvae); grape cane gallmakers; grape leafhoppers; grape mealybugs; grape phylloxera; Japanese beetles.

Diseases, see page 94
Anthracnose; botrytis fruit rot; Pierce's disease; powdery mildew.

▶ **'Lakemont Seedless'** *grapes range in color from white to a clear amber.*

Recommended Cultivars

'Black Monukka' Suitable for California, Arizona and the Southeast. Very dark red seedless cultivar.

'Himrod' A white seedless cultivar most appropriate for the eastern United States and southern Canada. Ripens in mid-August. Resistant to downy mildew. Zones 5–8.

'Interlaken Seedless' Very sweet, golden fruit for fresh use, drying or juice. Zones 5–8.

'Lakemont Seedless' Suitable for eastern conditions through the warmest sections of Zone 5. Amber colored. Very disease-resistant. Zones 5–8.

'Olivette Blanche' Green seedless cultivar suitable for California and Arizona. Zones 7–10.

'Thompson Seedless' The most common European cultivar grown in the United States. Grown for table grapes, wine and raisins. A good choice for California and Arizona. Zones 7–10.

Grapes, Muscadine

Muscadine grapes, native to the southeastern United States, are more resistant to pests and diseases than either European or American grapes. They require the long, hot summers of the South to ripen their loose clusters of large, luscious fruit.

Attention required

Bee/butterfly friendly ✗

Ornamental ✓

Container growing ✗

Yield

Ease of propagation

TYPE AND APPEARANCE
● Muscadine grape (*Vitis rotundifolia*)
Perennial vine.
Height 12–20 feet; spread 3–5 feet; Zones 7–10.
Flowers: Tiny, greenish, borne in inconspicuous clusters; in spring.
Leaves: Medium to bluish green, unlobed with a jagged margin, 4–6 inches long; turn yellow in fall.

PLANTING CONSIDERATIONS
When to plant: Early spring or fall.
Where to plant: Full sun.
Soil and feeding: Plant in well-prepared soil with abundant organic matter incorporated. See page 28. Feed with 1–1½ pounds of compost per foot of row in late winter each year.
Planting distance: 15–20 feet between plants; 15 feet between rows.

AFTER PLANTING
Cultivate well for several feet around plants to eliminate weeds during the first year and in the spring of the second. Mulch in midsummer during the second year with 6 inches of hay or straw or 4 inches of ground corncobs. Continue cultivating during the first half of the season and mulching during the second half throughout the plant's life.
Watering: Apply 4 gallons per square foot of root area every week to ten days until August. (As a rule of thumb, the root area will be one and a half to two times wider than the crown, or top part of the plant.) Do not wet leaves. Do not water during the end of the season.
Pollinators: Some cultivars require pollinators, so ask before you buy. Pollen is wind-borne, not transferred by insects. Site pollinators within 50 feet of plants they are to fertilize.

HARVESTING
Time: Late summer and fall.
Test for ripeness: Taste for sweetness when grapes are fully colored. Pick only when ripe, since grapes will not continue to ripen off the vine. Shake muscadines off the vine onto a cloth held under the loose cluster.
Years to bearing: 2–3.
Yields per bush/plant: 20–30 clusters per plant (8–16 quarts per plant).

PROPAGATION
Layer canes in midsummer (see page 69). Separate from the parent and plant the following spring.

PRUNING AND TRAINING
Plants fruit on one-year-old canes that develop from main "arms," or branches.
After planting: Prune to a single cane with two buds.
Routine pruning:
Muscadines should be pruned and trained to an arbor or to the Munson system (see page 64). Prune back the fruiting canes that develop from the main arms to about six buds. Remove these canes after harvest to allow new canes space to grow. After the vine is mature, cut out one arm of the vine every year and let a new shoot take its place. This allows the vine to renew itself gradually.

Mature plants: Prune and train for the Munson system or on an arbor.

Neglected plants: Cut out old wood, leaving only the trunk and arms required for training to the Munson system. Follow directions for that system in subsequent years.

PESTS, see page 85
Birds; grape berry moths (and larvae); grape cane gallmakers; grape leafhoppers; grape mealybugs; grape phylloxera; Japanese beetles.

DISEASES, see page 94
Anthracnose; black rot; botrytis fruit rot; Pierce's disease; powdery mildew.

 Muscadine grapes *ripening to luscious fruits.*

RECOMMENDED CULTIVARS

'Carlos' Self-fertile; disease-resistant; will tolerate warmest parts of Zone 6. Medium size; gold with a pink blush.
'Hunt' Dull black fruit ripens evenly; disease-resistant;
makes a good pollinator for other cultivars.
'Magoon' Self-fertile; reddish purple, medium-size grapes; very aromatic.
'Scuppernong' Green- or red-
bronze fruits; very sweet and juicy; requires a pollinator such as 'Hunt'.
'Southland' Recommended for Florida; self-fertile; large, purple, very sweet fruit.

Kiwis

The potential of these Chinese vines was not commercially recognized in the West until the 1960s, when the first kiwi orchards were planted. Now, as home gardeners discover how easy they are to grow, kiwis are becoming more and more popular.

Attention required

 ✓
Bee/butterfly friendly

 ✓
Ornamental

 ✗
Container growing

Yield

Ease of propagation

NOTE
Kiwis are pollinated by bees but are less attractive than clover, which blooms at the same time. Mow clover when kiwi vines flower to reduce the distraction. Commercial growers keep bees in their kiwi orchards to ensure good pollination.

TYPE AND APPEARANCE
● *Actinidia deliciosa*, the fuzzy kiwi, has cultivars with egg-sized, hairy-skinned fruit.
● *A. arguta* and *A. kolomikta*, hardy kiwis, are more cold-tolerant with small, smooth-skinned fruit.
Height 15–30 feet; spread 3–5 feet.
● *A. deliciosa* Zones 7–10.
● *A. arguta* and *A. kolomikta* Zones 4–9.
Deciduous perennial vines.
Flowers: Cream-colored, showy, fragrant, ¾–1½ inches wide; in midspring.
Leaves: Bronze or reddish when young, turning dark green; smooth, oval, 6–8 inches long.

PLANTING CONSIDERATIONS
When to plant: Spring or fall.
Where to plant: Full sun for fuzzy kiwis except in the warmer parts of Zones 9 and 10, where some afternoon shade is required. For hardy kiwis, provide a sheltered site or north-facing slope in partial shade to protect plants from winter injury.

Soil and feeding: Plant in well-drained, slightly acid (pH 5.0–6.5) soil. See page 28. Kiwis are heavy feeders; provide a balanced organic fertilizer, scratching it into the soil around each vine in early spring; top-dress with compost.
Planting distance: 10 feet between plants, 15 feet between rows.

AFTER PLANTING
Cultivate the soil around each vine for the first season without adding mulch. During the second season, apply mulch after midseason.
Watering: Roots can grow 12 feet deep and extend out 10 feet from the vine in sandy soils. In clay soils, root growth may only be 3–4 feet deep and about as wide. In either case, water deeply, applying about 4 gallons of water per square foot of root area every week through the season. Do not wet foliage or fruit.

Pollinators: Most cultivars require a male plant for pollination. Those that are self-fruitful may yield more with a male nearby.

HARVESTING
Time: Fall. Kiwis taste best when allowed to ripen fully on the vine, but gather them before the first frost, which will ruin the fruit.
Test for ripeness: You can pick kiwis when slightly underripe as long as the fruit gives slightly under pressure from your thumb. Kiwi fruits keep for up to three months in cold storage or refrigeration.
Years to bearing: 3.
Yields per plant: 5–10 gallons per female plant.

PROPAGATION
Take softwood cuttings from young growth during June to August. Remove all but the top two leaves, plant in damp sand, cover with supported plastic film and mist several times a day. Cuttings should root in two to four weeks.

PRUNING AND TRAINING
Plants fruit on the current season's growth, which emerges from wood grown the previous year.
After planting: Grow kiwis on a T-bar trellis (see page 62). Train each vine to a stake and allow it to grow up until it reaches the center wire of the

T-bar trellis. Prune off any laterals (side branches) that grow from this central trunk. Prune back the tip of the trunk and let two shoots grow out, tying them to the center wire so they grow in opposite directions. Tie the flowering laterals coming from these two "arms" so they grow vertically. Ideally, the laterals should be spaced 10–15 inches apart. Shoots from these laterals will bear flowers and fruit the following year.

Routine pruning: During the dormant period, remove the shoots on which fruit grew the previous season, together with any damaged, tangled or unwanted laterals. In spring, cut back the flowering laterals to four to six leaves beyond the last flower. If fruit set is particularly heavy, prune off some fruit. During summer, prune off excess laterals.

Pruning male plants: Train on a T-bar trellis, as described previously, but after flowering, prune off flowering laterals close to the main arm.

Mature plants: Remove flowering laterals after three seasons. You can create new arms with shoots originating near the trunk.

Neglected plants: Try to reestablish plant form and training. Prune off old wood, including the main arms if damaged, and let new shoots grow to take their place.

PESTS, see page 85
Leafroller caterpillars; soft scales.

DISEASES, see page 94
Botrytis blight; crown rot.

RECOMMENDED CULTIVARS

'Arctic Beauty' Said to be the most cold-hardy cultivar of *A. kolomikta*. Male plants have lovely variegated foliage in shades of pink, white and bright green. High-yielding. Zones 3–8.

'Blake' A self-pollinating cultivar of *A. deliciosa* with nicely textured fruit and good flavor. Zones 4–8.

'Chico' Often grown in California commercial plantings, it is a selection of 'Hayward'. Zones 7–10.

'Chico' (male) Used to pollinate 'Hayward' or 'Chico'.

'Hayward' By far the most commonly grown female cultivar of *A. deliciosa* because of its fine flavor and storage qualities.

'Issai' A self-pollinating *A. arguta* strain with small (1½ inches), green, smooth-skinned fruit. Sweeter than fruits of *A. deliciosa*, with a long fruiting season (July to first frost). Zones 4–9.

'Vincent' A cultivar of *A. deliciosa* that is well suited to the warmest areas because it requires only 100 chilling hours. Zones 6–9.

 'Hayward' *gives high yields of perfect fruit.*

Loganberries

Bred in Canada in the 1880s, the first loganberries had thorny canes and a very sharp flavor, which made them suitable only for cooking. Today, less-thorny strains are widely available and the berries are much sweeter than on the old-fashioned types.

Attention required

 ✓

Bee/butterfly friendly

 ✓

Ornamental

 ✗

Container growing

Yield

Ease of propagation

TYPE AND APPEARANCE
• Erect loganberry (*Rubus ursinus* var. *loganobaccus* cv.) Plants are deciduous, with perennial roots and erect biennial canes.
Height 4–6 feet; spread 3–5 feet; Zones 5–9.
Flowers: Light pink, five petaled, 1 inch wide; in spring.
Leaves: Dark green, with three leaflets; dull red in fall.

PLANTING CONSIDERATIONS
When to plant: Spring or fall.

Where to plant: Full sun. Protect from winter winds.
Soil and feeding: Plants grow best in most well-prepared soil enriched with organic matter. See page 28. Feed with 1–1½ pounds of compost per foot of row in late winter each year.
Planting distance: 5–6 feet between plants; 8 feet between rows.

AFTER PLANTING
Cultivate soil shallowly to eliminate weeds until midsummer. Mulch with 6 inches of straw or weed-free hay or 3 inches of rotted wood chips.
Watering: 2 gallons per square foot of root spread, applied weekly until berries begin to color; then 1 gallon per square foot applied weekly until the first frost. (The root area will be one and a half to two times wider than the top of the plant.)
Pollinators: None required.

◀ **Loganberries** *are raspberry-like in texture but with a sharper flavor.*

HARVESTING
Time: July and early August.
Test for ripeness: When gently pulled, the dark fruit should "fall" into your hands. Taste for ripeness.
Years to bearing: 1.
Yields per plant: 2–6 quarts.

PROPAGATION
Tip layer in summer (see page 69).

PRUNING AND TRAINING
Same as for blackberries (see page 107) or raspberries (see page 149). Plants fruit on one-year-old wood.

PESTS, see page 84
Aphids; raspberry red-necked cane borers; raspberry fruit worms; raspberry sawfly; spider mites.

DISEASES, see page 93
Anthracnose; botrytis fruit rot; crown gall; crumbly berry virus; mosaic virus; powdery mildew; Verticillium wilt.

RECOMMENDED CULTIVARS
Because loganberries are a cross between raspberries and blackberries, cultivars are not often identified. Buy locally to make sure the selection grows well in your yard; Zones 6–10. Taste fruit before buying plants to determine its sweetness or acidity.

Nectarines

Nectarines are favorites with anyone who loves peaches but doesn't love their fuzzy skins. If you can grow a peach, you can grow a nectarine because, aside from preferring warmer summer temperatures, they require the same treatment.

TYPE AND APPEARANCE
● Nectarine (*Prunus persica* var. *nucipersica*)
Deciduous tree.
Height 4–20 feet; spread 6–25 feet; Zones 5–9.
Flowers: Pink, double in some cultivars, highly decorative, 1–2 inches wide; in very early spring.
Leaves: Medium green with a light sheen, decorative, 6–8 inches long, in drooping clusters; turn yellow in fall.

PLANTING CONSIDERATIONS
When to plant: Early spring. In mild-winter areas, in spring or fall.
Where to plant: Site in full sun, in a protected niche. Where late-spring frosts are common, plant on a north-facing slope. Don't plant where peaches or nectarines were grown before.
Soil and feeding: Plant in well-drained, sandy loam with high fertility. See page 28. Apply 5–10 pounds of compost around each tree in late winter. Spread a cup of alfalfa meal and 2 cups of bonemeal

around each plant after petal drop.
Planting distance: At least the height of the mature tree — 8–12 feet apart for grafted dwarfs, 4 feet for genetic dwarfs and 15–20 feet apart for standard trees.
Rootstocks: Affect height, spread and vigor. See page 24.

AFTER PLANTING
Keep root area weed-free and mulch with 6–12 inches of straw or 3–6 inches of wood chips or other organic mulch that settles. A deep mulch is required for northern areas during winter.
Watering: Apply 2 gallons of water per week per square foot of root area. (As a rule of thumb, the root area will be one and a half to two times wider than the crown, or top part of the plant.)
Pollinators: None required for most cultivars.

HARVESTING
Time: Late July, August and early September.
Test for ripeness: Pick when

fully colored. Ripe fruit gives slightly when squeezed, and the fruit stem separates easily from tree. Taste for ripeness. If frost threatens, pick slightly early; fruit will continue to ripen after being picked.
Years to bearing: 3–5.
Yields per tree: 3–5 bushels (standard); 1–3 bushels (dwarf).

PROPAGATION
Graft in early spring or bud in late summer. See pages 70 and 71.

PRUNING AND TRAINING
Plants fruit on one-year-old wood.
After planting: Nectarines should be trained to an open-center or modified central-leader form. See page 101 in the Apples entry.
Routine pruning: Nectarines require a great deal of pruning each year to remain prolific, since fruit only forms on one-year-old wood. Early pruning stimulates early flowering, so wait until late spring or early summer for all pruning operations. Cut back the longest branches, including the modified leader, every year to retain manageable size. Thin fruit to 6–8 inches apart on standard trees, 5–7 inches apart on dwarfs.
Mature plants: Prune out weak, dead and crossing branches. Continue to prune back long growth.

Attention required

Bee/butterfly friendly

Ornamental

Container growing

Yield

Ease of propagation

NOTE
Chilling requirements range between 800 and 1,200 hours. Nectarines, like peaches and apricots, bloom before leafing out in the spring. Plant them in sheltered areas to protect against frost, or grow dwarf trees in containers that can be moved inside when necessary.

▲ **'Fantasia'** *bears sweet, juicy fruit.*

Neglected plants: Prune out all dead, weak and poorly positioned branches. Cut back long branches. Remember to thin fruit, particularly if you have removed a good deal of leaf area.

Training: Plants can also be trained as fans, cordons or espaliers. See pages 58 and 59.

PESTS, see page 81
Oriental fruit moth larvae; peachtree borers.

DISEASES, see page 90
Brown rot; peach scab.

RECOMMENDED CULTIVARS

'Fantasia' A bright red and yellow freestone fruit, appropriate for California. Fruits late in the season and is very flavorful. Zones 6–10.

'Hardired' As noted in the name, this nectarine is quite hardy. Fruit is relatively small, freestone and has red skin. Very firm flesh is good for canning or baked goods in addition to eating fresh. Tolerant of brown rot. Zones 5–9.

'June Glo' A patented cultivar that is especially good in the humid conditions of the Northwest. Productive. Red-skinned, flavorful fruit. Zones 6–10.

'Mericrest' A heavy yielding, freestone fruit with red skin. Ripens early. Relatively hardy. Good disease resistance. Zones 5–9.

'Pocahontas' Appropriate for the Southeast, this disease-resistant cultivar has medium-size fruits and large blossoms. Zones 6–10.

'Sunred' A cultivar with yellow skin overlaid with a red blush. Appropriate for Zone 9 conditions since it has a low chilling requirement. Zones 6–10.

Pawpaws

Grow a pawpaw tree for the kids. Children love fresh pawpaws and so do most adults. Fresh, they have a soft, custardy texture and taste a little bit like a ripe, sweet banana, with overtones of pineapple and mango.

TYPE AND APPEARANCE

● Pawpaw (*Asimina triloba*) Deciduous tree.
Height 20–30 feet; spread 10–15 feet; Zones 5–8.
Flowers: Maroon colored, unusual but attractive, 1–2 inches wide; in spring.
Leaves: Light green, large, drooping, 10–12 inches long; turn yellow in autumn.

PLANTING CONSIDERATIONS

When to plant: Spring; all season for container-grown plants.
Where to plant: Full sun.
Soil and feeding: Plant in well-drained, nutrient-rich soil. See page 28. Apply 5 pounds of compost around each plant in late winter each year.
Planting distance: 25 feet apart.

AFTER PLANTING

Keep root area weed-free and mulch with 6 inches of straw or leaves or 3–4 inches of wood chips.
Watering: Apply 2–3 gallons per week per square foot of root area (one and a half to two times wider than the top part of the plant).
Pollinators: Most pawpaws need a pollinator, so buy two different cultivars. For heaviest fruit set, hand-pollinate.

HARVESTING

Time: August and September.
Test for ripeness: Harvest when fully colored and slightly soft. Fruit will continue to ripen after they are picked.
Years to bearing: 4–6.
Yields per tree: 1–3 bushels.

PROPAGATION

Propagate by taking suckers from nongrafted cultivars or by grafting from grafted trees. See pages 68 and 70.

PRUNING

Plants fruit on one-year-old wood.
After planting: Reduce top growth by up to a third to balance possible root loss. Train grafted pawpaws as central-leader trees. Own-root plants can be left to sucker at will and become shrubs.

Routine pruning: Cut out dead, weak or crossed branches. Prune central-leader trees to a pyramid shape. Thin bushy plants by removing weakest or thinnest branches.
Mature plants: Shorten shoots to promote fruitful new growth.
Neglected plants: Prune out dead and weak wood; reshape as desired.

PESTS AND DISEASES

Generally pest- and disease-free if sanitation is good.

RECOMMENDED CULTIVARS

You'll get the largest, most flavorful fruit if you buy named cultivars.
'Overleese' Large, oval to round green fruits with excellent flavor. Zones **6–9**.

▲ **Harvest** *when fruits are dark brown, almost black.*

Attention required

✓ Bee/butterfly friendly

✓ Ornamental

✗ Container-growing

Yield

Ease of propagation

NOTE
Pawpaws are notoriously difficult to transplant because of a long and tender taproot. Buy only container-grown plants and disturb the root system as little as possible when planting. Keep well watered and you should have no trouble.

Peaches

There are few taste experiences more heavenly than eating a sun-ripened peach straight off the tree. Prolong your picking season by planting several cultivars, but choose cultivars appropriate to your climate — peaches are environmentally demanding fruits.

Attention required

Bee/butterfly friendly   ✓

Ornamental   ✓

Container growing   ✓

Yield

Ease of propagation

NOTE
Chilling requirements range from 200–1,200 chilling hours. If you live in a mild-winter area, choose a low-chill cultivar, but if your winters are cold, select a high-chill cultivar. Check with your supplier for each cultivar's needs.

TYPE AND APPEARANCE
● Peach (*Prunus persica*) Deciduous tree. Height 15–20 feet (standard), 4–10 feet (dwarf); spread 20–25 feet (standard), 6–15 feet (dwarf); Zones 5–9.
Flowers: Pink to almost red, 1–2 inches wide, decorative; in very early spring.
Leaves: Glossy green, narrow, about 3–6 inches long, in clusters.

PLANTING CONSIDERATIONS
When to plant: Early spring; spring or fall in mild-winter areas.
Where to plant: Site in full sun. Where late-spring frosts are common, plant on a north-facing slope. Don't plant where peaches or nectarines were grown before.
Soil and feeding: Plant in sandy loam that is well drained, with high organic matter and fertility. See page 28. Apply 5–10 pounds of compost around each tree in late winter. Spread a cup of alfalfa meal and 2 cups of bonemeal around each plant after petal fall.

Planting distance: 15–20 feet (standard); 8–12 feet (grafted dwarf); 4 feet (genetic dwarf).
Rootstocks: Affect height, spread and vigor. See page 24.

AFTER PLANTING
Keep root area weed-free and mulch with 6 inches of straw or 3–4 inches of denser material such as wood chips. Double the depth of mulch over winter in cold climates. Protect plants from sunscald by painting the trunks from the soil to the first branch with diluted white latex paint.
Watering: Apply 2–3 gallons of water per week per square foot of root area. Do not let the soil dry out at any time.
Pollinators: None required for most cultivars.

HARVESTING
Time: Late July through August and September, depending on the cultivar.
Test for ripeness: Peaches taste best when allowed to ripen fully on the tree. Check the skin color under twig

attachment: Skin that has turned from green to yellow indicates ripeness. If frost threatens, harvest full-size and well-colored fruit even if slightly underripe. It will ripen rapidly at room temperature.
Years to bearing: 4–5.
Yields per tree: 2–3 bushels (standard); 1–2 (dwarf).

PROPAGATION
Graft in early spring or bud in summer. See pages 70 and 71.

PRUNING AND TRAINING
Plants fruit on one-year-old wood.
After planting: Prune to either an open-center or modified central-leader system. See the Apples entry

CONTAINER GROWING
Choose genetic dwarfs or trees on dwarfing rootstocks for container growing. Use a container at least 1½ feet deep and wide. Once your plant reaches full size, repot it every year just before growth begins in spring, and prune back the roots and top. In very cold areas, move the plant under cover for the winter. If this is not possible, protect the roots in winter by encasing the pot in chicken wire, and filling with leaves and wrapping as described for Figs on page 122.

From *Growing Fruits and Vegetables Organically* (Rodale Press 1994)

on page 101. Prune plants trained to a central leader to 30 inches above the soil surface for an open system and to 36 inches for a modified central leader.

Routine pruning: Peaches require a great deal of annual pruning, since good fruit production depends on stimulating new wood that will fruit the following year. Cut back leading branches to retain desired size, and cut back laterals to about two-thirds of their length. Postpone pruning until after blossoming. Thin fruit to a distance of 6–8 inches apart.

Mature plants: Cut back all dead, weak or poorly positioned branches. Remove two-thirds of the laterals that bore fruit the previous year. Cut back all long growth.

Neglected plants: Prune off all dead, weak or poorly positioned branches. Remove two-thirds of all branches that have recently borne fruit. Cut back leading branches and side branches.

Training: Plants can also be trained as fans, cordons or espaliers. See pages 58 and 59.

PESTS, see pages 81 and 82 Oriental fruit moth; plum cucurlio; scales; spider mites.

DISEASES, see pages 90 and 91 Bacterial leaf spot; peach scab; peach leaf curl.

RECOMMENDED CULTIVARS
'Desert Gold' A cultivar for Arizona and California because of its low chilling requirements and good commercial qualities including fine flavor. This semi-freestone requires heavy fruit thinning. Zones 7–10.

'Frost' A flavorful cultivar for the Northwest. Medium-size, semi-freestone fruit with good fresh eating and preserving characteristics. Resistant to leaf curl disease. Zones 6–9.

'Honey Babe' A genetic dwarf with large, flavorful fruit. Tree can be container grown, since it grows only 3–5 feet tall. Zones 7–9.

'Redhaven' A hardy peach, suitable for protected Zone 5 conditions. Red-skinned, medium-size, semi-freestone fruits have excellent flavor. 'Early Redhaven' ripens somewhat earlier but is similar in other respects. Zones 5–9.

'Reliance' A very hardy peach that can withstand −25°F during midwinter. Large fruit is freestone, red skinned and firm enough to freeze and can well. Some disease resistance. Zones 4–9.

'Rio Oso Gem' A cultivar appropriate to the South and West. Flowers relatively late. Red over yellow freestone fruit is good for both fresh eating and preserving. Zones 5–8.

◄ **'Frost'** *is a good choice for the Pacific Northwest.*

Pears

As a home gardener, you are lucky — you can choose pear cultivars with juicy fruit and wonderful flavor rarely available at the grocery store. Pears have a wide range of culinary uses, and most cultivars store well for several months in cold, dark conditions.

Attention required

Bee/butterfly friendly ✓

Ornamental ✓

Container growing ✓

Yield

Ease of propagation

NOTE
Most cultivars require 900–1,000 chilling hours.

TYPE AND APPEARANCE
● European pear (*Pyrus communis*)
Deciduous tree.
Height 25–40 feet (standard), 15–20 feet (semidwarf), 8–15 feet (dwarf); spread 25 feet (standard), 15 feet (semidwarf), 10–15 feet (dwarf); Zones 4–9.
Flowers: White, in decorative clusters, 1–1½ inches wide; in spring.
Leaves: Dark green, slightly toothed margins, 1–2½ inches long; turn red or yellow in fall.

PLANTING CONSIDERATIONS
When to plant: Early spring or fall.
Where to plant: Plant in full sun, and choose a site with good air circulation.
Soil and feeding: Pears prefer well-drained soil, but can tolerate heavy clay soil. They grow best in soil with a pH of 6.0–6.5. See page 28. Don't overfertilize pears — too much nitrogen can promote fire blight. Apply 5–10 pounds of compost around each tree in late winter.

Planting distance: 25–30 feet (standard); 15–20 feet (semidwarf); 15 feet (dwarf).
Rootstocks: Affect height, spread and vigor. For fire blight resistance, choose pears with rootstocks in the 'Old Home' × 'Farmington' ('OH × F') series, which range from full size to very dwarfing. See page 24.

AFTER PLANTING
Keep the root area weed-free and mulch with 6 inches of straw or 3–4 inches of denser material, such as wood chips.
Watering: Apply 2–3 gallons of water per square foot of root area weekly. (The root area will be one and a half to two times wider than the top part of the plant.)
Pollinators: Check with your nursery or supplier to choose the best pollinators for your plants. 'Seckel' and 'Bartlett' do not pollinate each other, and 'Magness' does not pollinate any cultivar.

HARVESTING
Time: August to October, depending on the cultivar.
Test for ripeness: Do not let pears ripen on the tree, because their texture will deteriorate. Instead, pick them when full size but before they have softened. Ripen pears indoors or store them in the refrigerator until a week before you plan to eat them. You'll get faster ripening if you place pears inside a plastic bag with a ripe apple.
Years to bearing: 4–7.
Yields per tree: 2–4 bushels (standard); ½–1½ bushels (dwarf).

PROPAGATION
Graft in early spring or bud in summer. See pages 70 and 71.

PRUNING AND TRAINING
Plants fruit on one- to ten-year-old spurs and terminals (tips) of laterals.
After planting: Train to a central leader or modified central leader. See the Apples entry on page 101.
Routine pruning: For heavier fruiting, counteract the tree's tendency toward vertical growth by spreading developing branches with clothespins, toothpicks or wooden bars. If a cultivar is not resistant to fire blight, leave up to eight scaffold branches, rather than the customary four to six, in case

◄ **'Comice'** *has an excellent texture and fine flavor.*

some must be removed. Do not prune off the tips of all the fruiting shoots on cultivars that fruit on terminal buds of shoots as well as spurs. Terminal buds on shoots bloom later than those on spurs and are less likely to be killed by a late frost.

Most cultivars do not require much fruit thinning. However, young trees with high fruit set should be thinned to 6 inches between fruits.

Mature plants: Thin new laterals annually and remove dead, weak or poorly positioned branches. Prune out old wood when spurs are seven or eight years old.

Neglected plants: Prune dead, weak or poorly positioned branches. Do not cut back laterals severely, or too much new, disease-susceptible growth will be stimulated.

Training: Plants can also be trained as fans, cordons or espaliers. See pages 58 and 59.

PESTS, see pages 80 and 82 Codling moth; pear psylla.

DISEASES, see page 91 Fire blight; pear scab.

RECOMMENDED CULTIVARS

'Bosc' Fruits with excellent flavor that grow singly, and keep up to four months at a temperature of 35°F. Best for Zones 6–9, though it usually fruits in protected spots in Zone 5.

'Comice' One of the finest-tasting pears. Does not fruit or ripen well in conditions cooler than Zone 6. Widely grown in the Northwest and California. Ripens in October through most of its range. Zones 6–9.

'Highland' Grown in the Northwest. Excellent flavor, texture and keeping qualities. Tastes best after about a month in cold storage. Yellow skin with some russeting on large fruit. Zones 5–8.

'Mericourt' White-fleshed with yellowish skin blushed with red. Resistant to fire blight and leaf spot. Appropriate for the Southeast. Zones 6–9.

'Seckel' Small with rough brown skin. Wonderful flavor and texture, and can be picked ripe. Grows well in most parts of the United States and Canada. Fire blight resistant. Zones 3–8.

'Ure' Hardy in protected spots in Zones 3 and 4. Small fruit with thin greenish yellow skin and buttery and aromatic flesh. Zones 3–8.

Pears, Asian

Asian pears are crisp and juicy with a flavor all their own. They require warmer temperatures than European cultivars and are more resistant to pests and diseases such as pear psylla and pear scab. Store them between four and eight months in cold conditions.

Attention required

Bee/butterfly friendly ✓

Ornamental ✓

Container growing ✗

Yield

Ease of propagation

NOTE
Chilling requirements are only 400 to 900 hours, causing trees to bloom very early.

TYPE AND APPEARANCE
● Asian pear (*Pyrus pyrifolia, P. ussuriensis* and *P. bretschneideris* hybrids) Deciduous tree.
Height 25–30 feet (standard), 15 feet (semidwarf), 8–12 feet (dwarf); spread 25 feet (standard), 15 feet (semidwarf), 10–15 feet (dwarf); Zones 5–9.
Flowers: White, very showy, 1–1½ inches wide; in spring.
Leaves: Dark green, slightly toothed margins, 1–2½ inches long; turn red or yellow in fall.

PLANTING CONSIDERATIONS
When to plant: Early spring or fall.
Where to plant: Plant in full sun, and choose a site with good air circulation.
Soil and feeding: Plants can tolerate heavier soil than most fruits, but good drainage promotes good health. See page 28. Don't overfertilize — too much nitrogen promotes diseases such as fire blight. Apply 5–10 pounds of compost around each tree in late winter.

Planting distance: 25 feet (standard); 15 feet (semidwarf); 8–10 feet (dwarf).
Rootstocks: Affect height, spread and vigor. See page 24. An 'Old Home' × 'Farmingdale 513' rootstock produces a semidwarf tree with better cold-hardiness plus more resistance to fire blight and tolerance of wet soils.

AFTER PLANTING
Mulch with 6 inches of straw or 3–4 inches of denser material, such as wood chips.
Watering: Apply 2–3 gallons of water per week per square foot of root area. (As a rule of thumb, the root area will be one and a half to two times wider than the crown, or top part of the plant.)
Pollinators: Check with your nursery or supplier for an appropriate pollinator for your cultivars.

HARVESTING
Time: August, September and early October.
Test for ripeness: Let fruit ripen on the tree to develop the best texture and flavor. Taste to determine ripeness after fruit has reached full size and color.
Years to bearing: 4–7.
Yields per tree: 3–8 bushels (standard); 1–2 bushels (semidwarf); ½–1 bushel (dwarf).

PROPAGATION
Graft in early spring or bud in summer. See pages 70 and 71.

PRUNING AND TRAINING
Plants fruit on one- to six-year-old spurs.
After planting: Train dwarf trees to a modified central leader and standards to an open center. See the Apples entry on page 101. Head back the central leader to 36 inches above the soil surface. Trim topgrowth to balance root loss.
Routine pruning: As described for pears (see page 138), with the difference that Asian pear cultivars may need more annual pruning since they tend to grow more vigorously. Thin fruit to only one per spur to promote the best plant health and largest fruit size.
Mature plants: Same as for pears; see page 139.

Neglected plants: Same as for pears; see page 139.
Training: Plants can also be trained as fans, cordons or espaliers. See pages 58 and 59.

PESTS, see page 82
Aphids; pear slugs.

DISEASES, see page 91
Fabraea leaf spot; pseudomonas leaf blight.

RECOMMENDED CULTIVARS
'Dan Bae' A cultivar from Korea. Olive green fruits keep till spring, even if not refrigerated. Blooms early and ripens late. Appropriate for the Pacific Northwest and coastal California. Zones 5–9.
'Hosui' A cultivar that has brownish skin with rose, yellow and green undertones. Medium to large fruit with excellent flavor. Ripens in September. Tender tree that is appropriate for Zones 7–9.
'Seuri' A fire blight-resistant cultivar with large orange fruit. Late ripening. Zones 6–9.
'Shinko' A good keeper. Fruit is bronze, russeted and flavorful. Medium size and late maturing. Zones 5–9.
'Shinseiho' A cultivar known for its large size and good flavor and texture. Fruit has yellow-green skin. The flavor is both sweet and tart. Widely adaptable. Zones 6–9.
'20th Century' A cultivar grown in the Northeast as well as climates kinder to pears. Resistant to pear psylla. Fruit is medium size with clear yellow skin. Highly productive. Zones 5–8.

◀ **Wait** *until fruits are full size before picking.*

Persimmons

To enjoy the best flavor, allow persimmons to ripen at room temperature before eating, as most cultivars taste astringent before they soften. Unsoftened fruit will store for up to two months at 32°F with high humidity.

Attention required

Bee/butterfly friendly ✓

Ornamental ✓

Container growing ✗

Yield

Ease of propagation

NOTE
As persimmons grow very slowly, they make excellent espalier plants.

TYPE AND APPEARANCE
- American persimmon (*Diospyros virginiana*) Height 30–40 feet; spread 30 feet; Zones 5–9.
- Oriental persimmon (*Diospyros kaki*) Height 25–30 feet; spread 25 feet; Zones 7–10. Deciduous trees.

Flowers: Waxy white, fragrant, ½–2 inches wide; in summer.

Leaves: American persimmon — dark green, glossy, 6 inches long; turn yellow in fall. Oriental persimmon — dark green, glossy, heart shaped, 5–7 inches wide; turn brilliant orange or yellow in fall.

PLANTING CONSIDERATIONS
When to plant: Spring for bareroot plants; any time during the growing season for container-grown plants. Persimmons have a long taproot, so dig extra-deep holes for them and transplant carefully.

Where to plant: Full sun.
Soil and feeding: Plant in well-drained soil. Oriental cultivars prefer a moderately fertile sandy loam, while American cultivars will tolerate a broader range of soil types. See page 28. High-nitrogen soils promote fruit drop, so don't overfertilize. Apply 5–10 pounds of compost around each tree in late winter.

Planting distance: Leave enough space to ensure leaves of mature trees do not touch — 25–30 feet.

AFTER PLANTING
Keep root area weed-free. Mulch with 6 inches of straw or 3–4 inches of denser material such as wood chips.
Watering: Apply 2–3 gallons per square foot of root area during the first season. In subsequent years, persimmons require very little watering. In dry conditions, supply 2–3 gallons per square foot of root area every two weeks. (As a rule of thumb, the root area will be one and a half to two

times wider than the crown, or top part of the plant.)
Pollinators: Required by some Oriental cultivars and most American cultivars. Check with your nursery.

HARVESTING
Time: September to October.
Test for ripeness: When fully colored and beginning to soften, clip off fruit of Oriental persimmons with stem attached. Harvest American persimmons when they are ripe enough to drop from the trees. Frozen fruit can be left on the tree and gathered as required in the early winter. Let all fruit soften before eating, unless you've chosen an Oriental cultivar that is nonastringent and can be eaten crisp.
Years to bearing: 2–3.
Yields per tree: 1 bushel (American); 1–2 bushels (Oriental).

▼ **Persimmons** *should be left to ripen at room temperature.*

▶ **'Hachiya'** *prefers California conditions but does well in other warm areas through Zone 6.*

PROPAGATION
Graft in early spring for Orientals and most named American cultivars. See pages 70 and 71. Ungrafted American types can be propagated from suckers. See page 68.

PRUNING AND TRAINING
Plants fruit on one-year-old wood.

After planting: Train as for a central leader. See the Apples entry on page 101.

Routine pruning: Leave six to eight scaffold branches, placed evenly around trunk. Remove excess growth and any dead or poorly positioned wood. To retain tree shape, remove suckers, which form readily on American cultivars.

Mature plants: Very little pruning required. Because fruit forms only on one-year-old wood, thin old fruiting branches when they become crowded.

Neglected plants: Thin out dead, weak and poorly positioned branches and old wood.

Training: Plants can also be trained as fans, cordons or espaliers. See pages 58 and 59.

PESTS, see page 81
Scales.

DISEASES, see page 91
Anthracnose.

RECOMMENDED CULTIVARS

'Fuyu' Small tree that requires a pollinator. Fruit can be eaten firm as well as softened. Sometimes called 'Jiro'. Zones 6–9.

'Gailey' Often grown to pollinate other Oriental cultivars. Zones 6–9.

'Hachiya' An Oriental cultivar that is popular in California. Very sweet when fully ripened, but astringent until then. Almost seedless. Requires a pollinator. Zones 6–9.

'John Rick' An American cultivar with excellent flavor. Requires a pollinator. Zones 5–8.

'Meader' An American cultivar, it is almost seedless and self-fruitful but can profit from a nearby male tree. Hardy to −30°F. Zones 3–8.

'Tanenashi' An Oriental cultivar grown in the Southeast. Can bear fruit without pollination. Large fruits of excellent quality. Zones 6–9.

Plums

Attention required

  ✓

Bee/butterfly friendly

  ✓

Ornamental

  ✓

Container growing

Yield

Ease of propagation

NOTE
Japanese plums generally require 700 to 1,000 chilling hours. European plums generally require 800 to 1,100 chilling hours.

With 15 species and around 2,000 cultivars, there's a plum for every taste. Most are either of European or Japanese origin. European plums are sweet and oval, while Japanese plums are larger, juicier and never have blue or purple skins.

TYPE AND APPEARANCE
● European plum (*Prunus domestica*)
Height 15–20 feet (standard), 1–12 feet (dwarf); spread 15–20 feet (standard), 10–15 feet (dwarf); Zones 4–9.
● Japanese plum (*Prunus salicina*)
Height 15–18 feet (standard), 8–12 feet (dwarf); spread 15–20 feet (standard), 10–15 feet (dwarf); Zones 6–10. Deciduous tree.
Flowers: White or pink, very decorative, ¾–1 inch wide; in very early to early spring.
Leaves: Medium green or purplish, sawtoothed margins, 2–4 inches long; turn yellow in fall.

PLANTING REQUIREMENTS
When to plant: Early spring; spring or fall in mild-winter areas.
Where to plant: Plant in full sun. A north-facing slope is ideal.
Soil and feeding: Plant in well-drained soil with high nutrient content. European plums will tolerate somewhat

heavy clay soil, while Japanese cultivars grow best in sandy loam soil. All plums thrive in raised beds. See pages 17 and 28. Apply 5–10 pounds of compost around each tree in later winter. After petals drop, apply 1 cup of alfalfa meal and 2 cups of bonemeal per tree.
Planting distance: 25 feet (standard), 15 feet (dwarf).
Rootstocks: Affect height, spread and vigor. 'Nemaguard' resists nematodes, 'Mariana' increases hardiness, and 'Damas', 'Mariana' and 'Myrobalan' are good choices for wet or clay soils. See page 24.

AFTER PLANTING: Keep root area weed-free. Mulch with 6 inches of straw or 3–4 inches of denser material such as wood chips.
Watering: Apply 2–3 gallons of water per week per square foot of root area. (As a rule of thumb, the root area will be one and a half to two times wider than the crown, or top part of the plant.) Do not let plums experience water stress.

Pollinators: Some European plums require a pollinator, and most Japanese plums require a pollinator. Check with your nursery or supplier to find the best pollinators for your cultivars.

HARVESTING
Time: July to September.
Test for ripeness: Pick when fruit is fully colored. Taste to test for sweetness.
Years to bearing: 3–4.
Yields per tree: 1–2 bushels (standard); ½–1 bushel (dwarf).

PROPAGATION
Graft in early spring or bud in summer. See pages 70 and 71.

PRUNING AND TRAINING
European cultivars fruit on two- to six-year-old spurs; Japanese cultivars fruit on one-year-old wood and older spurs.
After planting: Cut back whips (single stems) to 24–30 inches. Prune and train spreading cultivars (most Japanese cultivars) to an open-center form. See the Apples entry on page 101.
  Prune and train European cultivars to a modified central leader form. See the Apples entry on page 101.
Routine pruning: European cultivars require little pruning after the basic framework has been formed. Thin suckers and

water sprouts. Remove dead, weak or poorly positioned branches. Thin fruit so no more than two form on each spur.

Japanese cultivars require more annual pruning to retain desired size and shape. Cut back overly long branches. Remove excess branches and all weak, dead or poorly positioned growth. Thin fruit to 4–6 inches apart.

Mature plants: Prune European cultivars as little as possible, taking care to remove suckers and water sprouts. Prune Japanese cultivars to retain open center, good air circulation and desired size. It may be necessary to prune back laterals.

Neglected plants: Prune out all growth interfering with the basic shape of the tree. Prune out dead, weak or crossing branches. Remove suckers and water sprouts. Cut back overly long branches.

Training: Plants can also be trained as fans, cordons or espaliers. See pages 58 and 59.

PESTS, see page 82
Aphids; plum cucurlio; red spider mites.

DISEASES, see pages 90 and 91
Black knot; brown rot.

▼ **'Damson'** *is best known for its excellent drying and cooking qualities.*

RECOMMENDED CULTIVARS

'Damson' A dark purple plum that is a cultivar of *Prunus insititia*. Small, tart fruit that makes excellent preserves. Resists leaf spot. Hardy in Zones 4–9.

'Green Gage' A European cultivar with green skin and yellow flesh that ripens in September. Self-fruitful. Resists leaf spot. Hardy in Zones 4–10.

'Hybrid Kaga' A Japanese hybrid with red skin, yellow flesh and very sweet flavor. Hardy in Zones 5–9.

'Methley' A July-ripening Japanese hybrid. Very sweet, medium-size fruit with reddish purple skin and red flesh. Self-fertile; appropriate for the Northwest. Zones 6–8.

'Mount Royal' A European plum hardy even in relatively warm, protected spots in Zone 3. Yellow flesh and clingstone, with blue skin. Ripens in August to September. Zones 3–8.

'Santa Rosa' A very popular Japanese plum with red skin and flesh that shades from purple-red near the skin to pinkish yellow near the pit. Strains grow well from northern California to southern New England. Ripens in June in California and in July farther north. Resists black knot. Zones 5–9.

Plumcots (*P. simonii*) are featured by some nurseries. These hybrids, with apricot and Japanese plum parentage, taste rather like an apricot.

Quinces

Never eat a raw quince! Instead, make some tasty preserves or jelly, or stew some up with apples for a spectacular "quin-ap-sauce" or "quin-ap" pie. Cooked, quinces are delicious, but raw, they are hard, sour and astringent. The small, spreading tree forms a lovely accent plant.

Attention required

 ✓
Bee/butterfly friendly

 ✓
Ornamental

 ✗
Container growing

Yield

Ease of propagation

TYPE AND APPEARANCE
• Quince (*Cydonia oblonga*) Deciduous tree.
Height 10–15 feet; spread 10–15 feet; Zones 5–9.
Flowers: White or pink, very decorative, 2 inches wide, on tips of branches; in midspring to late spring.
Leaves: Heavy, oblong, with woolly undersides, 4 inches long; turn yellow in fall.

PLANTING CONSIDERATIONS
When to plant: Spring or fall.
Where to plant: Full sun.
Soil and feeding: Plant in well-drained, fertile, moisture-retentive soil. See page 28. Apply 5 pounds of compost around each tree or shrub in late winter.
Planting distance: 15 feet apart.

AFTER PLANTING
Keep root area weed-free. Mulch with 6 inches of straw or 3–4 inches of denser material such as wood chips.
Watering: Apply 2 gallons of water per week per square foot of root area. (As a rule of thumb, the root area will be one and a half to two times wider than the crown, or top part of the plant.)
Pollinators: None required.

HARVESTING
Time: September and October.
Test for ripeness: Fruit can be picked when it has turned yellow and is fragrant.
Years to bearing: 3–4.
Yields per tree: 1 bushel.

PROPAGATION
Bud in summer or take hardwood cuttings in fall. See page 71.

PRUNING AND TRAINING
Plants fruit on tips of last year's growth.
After planting: Decide whether you want to prune to a bush or an open-center tree. If pruning to bush form, cut off the leader about 1 foot above the ground and allow suckers to grow. If growing as a tree, develop the basic open-center framework, as described in the Apples entry on page 101, and cut back the central leader to 30–36 inches.

Routine pruning: Prune as little as possible to avoid fire blight infection. Do not thin fruit. Maintain desired shape of tree.
Mature plants: Prune off dead, weak and poorly positioned branches. Cut back main branches every few years to stimulate the formation of new shoots. Do not overprune.
Neglected plants: Prune out dead, weak or poorly positioned branches. Thin lateral branches to let air circulate freely within tree canopy.
Training: Plants can also be trained as fans, cordons or espaliers. See pages 58 and 59.

PESTS, see pages 80 and 82
Codling moth; oriental fruit moth.

DISEASES, see page 91
Fire blight; pseudomonas leaf blight.

▲ **Quinces** *are extremely ornamental in bloom.*

▲ **'Grand Champion'** *adds a special flavor to apple pies and sauces.*

RECOMMENDED CULTIVARS

Check with your local nursery to find out which cultivars do particularly well in your climate. If possible, buy a named and vegetatively propagated cultivar rather than a seedling tree — you'll get better flavor and more reliable hardiness.

'Grand Champion' is known for its high-quality fruit. Zones 5–8.
'Orange' A cultivar with orange-yellow skin that ripens early. Zones 5–8.
'Pineapple' Named for its flavor, not its appearance. The round fruits are excellent in jellies and sauces. Zones 5–8.
'Smyrna' The most popular cultivar because of its excellent, sweet flavor when cooked. Large, yellow, wonderfully fragrant oblong fruit. Zones 5–8.

Raspberries

Red raspberries are the bramble of choice for northern gardeners. Red and golden cultivars are more winter-hardy than the trailing black and purple cultivars. Choose from midsummer-bearing standards or everbearers, which fruit until fall.

TYPE AND APPEARANCE
- Red raspberry (*Rubus idaeus*)
- Black raspberry (*Rubus occidentalis*)

Other *Rubus* species such as *R. arcticus*, *R. kuntzeanus* and *R. parvifolius* are also used to breed cultivars with desirable characteristics.

Plants are deciduous, with perennial roots and biennial canes. Erect and trailing, summer-bearing and everbearing.

Height 6–15 feet; spread 4–5 feet; Zones 3–9.

Flowers: White or pink, five petaled, 1 inch wide; in spring.

Leaves: Dark green, with three leaflets; dull red in fall.

PLANTING CONSIDERATIONS
When to plant: Spring or fall.

Where to plant: Full sun. Protect purple and black cultivars from cold winter winds.

Soil and feeding: Plants grow best in well-prepared soil enriched with organic matter. See page 28. Feed with 1–1½ pounds of compost per foot of row in late winter each year.

Planting distance: 3–5 feet between plants; 5 feet between rows.

AFTER PLANTING
Cultivate surrounding soil shallowly to eliminate weeds until midsummer. Then mulch with 6 inches of straw or 3–4 inches of sawdust or wood chips.

Watering: Apply 2 gallons per square foot of root spread every week until developed berries begin to color; then apply 1 gallon per square foot of root spread every week until first frost. (As a rule of thumb, the root area will be one and a half to two times wider than the crown, or top part of the plant.)

Pollinators: None required.

HARVESTING
Time: Fruit on standards ripens in July and early August. The harvesting of everbearers depends on pruning; depending on how you prune, you can harvest in both midsummer and early fall or wait for a larger harvest in early fall.

Test for ripeness: Fruit should fall into your hands when pulled. Taste for ripeness.

Years to bearing: 1.

Yields per plant: 2–6 quarts.

PROPAGATION
Only propagate from disease-free plants.

Erect types: Dig up suckers in early spring and transplant. See page 68.

Trailing types: Tip layer in summer; see page 69. Transplant in fall in the South or the following spring in the North.

▼ **'Fall Gold'** *is sweet and succulent.*

Attention required

Bee/butterfly friendly ✓

Ornamental ✓

Container growing ✗

Yield

Ease of propagation

NOTE
Unlike other brambles, raspberries have hollow fruit—the "berry" separates from its receptacle when ripe. Black and purple raspberries do not yield as prolifically as red and yellow cultivars.

▲ **'Autumn Bliss'** *produces very large fruit.*

PRUNING AND TRAINING

Plants fruit on one-year-old wood or, if everbearing, in fall on the current year's growth.

After planting: Prune canes back to 6 inches tall.

Routine pruning—summer bearing: Head back primocanes (new canes) to 30 inches tall in midsummer. Prune out fruiting canes after harvest. In late spring, head back laterals on floricanes (flowering or second-year canes) to 18 inches long. Prune out weak canes.

Routine pruning— everbearing: Prune plants back to 6 inches tall after planting. In fall, some cultivars will fruit on the tips of these young canes. You can either leave these canes to overwinter or cut them off several inches above ground level. The following year, cut out only second-year canes after they have fruited in midsummer. Alternatively, cut down all canes in fall.

Mature plants: Continue with the routine pruning program throughout the life of the plant.

Neglected plants: Prune out all dead wood, prune back weak primocanes and follow routine pruning instructions.

Training: See pages 61 to 63.

PESTS, see page 84
Aphids; raspberry red-necked cane borers; raspberry fruitworms; raspberry sawfly; spider mites.

DISEASES, see page 93
Anthracnose; botrytis fruit rot; crown gall; crumbly berry virus; mosaic virus; powdery mildew; spur blight; Verticillium wilt.

RECOMMENDED CULTIVARS
'Autumn Bliss' A mosaic-resistant, everbearing red cultivar from England with fine flavor and large fruit size. Zones 4–8.

'Bababerry' An everbearing red cultivar that grows well all the way from Zones 4–10. Disease-resistant and productive, with good flavor.

'Canby' Thornless red, very round fruit; grows well in Zones 3–8.

'Fall Gold' A vigorous everbearing golden cultivar with good fruit flavor; Zones 4–9.

'John Robertson' A black cultivar hardy to Zones 4–8. Good flavor.

'Royalty' This purple summer bearer is unattractive to aphids. Can be harvested at red, purple and almost black stages. Zones 4–8.

Strawberries

Everyone loves strawberries and, fortunately for backyard gardeners, they are easy to grow. Plant early-, mid- and late-season cultivars to enjoy a plentiful supply of fresh berries. Fifty plants will allow a family unlimited summertime feasting and winter preserves.

Attention required

Bee/butterfly friendly ✓

Ornamental ✓

Container growing ✓

Yield

Ease of propagation

NOTE
In the warmest parts of the country, many gardeners plant strawberries in the fall and till or dig them under after they fruit the following spring.

TYPE AND APPEARANCE
● Alpine strawberry
(*Fragaria vesca*)
● June-bearing and day-neutral strawberries
(*Fragaria × ananassa*)
Herbaceous perennial.
Height 6–12 inches; spread 8–12 inches; Zones 3–10.
Flowers: White with prominent yellow center; five-petaled, 1 inch wide, in clusters on a stem arising from the crown; in early spring.
Leaves: Medium green, sometimes with red undertones; three leaflets with toothed edges; turn yellow, orange or scarlet in fall.
   Alpine strawberries have smaller flowers, leaves and fruit than June-bearing and day-neutral strawberries.

PLANTING CONSIDERATIONS
When to plant: In cold-winter areas, plant in early spring; in mild-winter areas, plant in late winter or fall.
Where to plant: Full sun for June-bearers and day-neutrals. Alpines can take some filtered afternoon shade. Plant on

sloping or higher ground to avoid frost pockets.
Soil and feeding: Plant in well-drained, fertile soil that's high in organic matter, with a pH of 5.5–6.5. See page 28.
   Avoid grubs by planting where no grass has grown for a year or two. Avoid some diseases by planting where no nightshade (tomato family) plant has grown for three years. Feed strawberries in

early summer, applying 2 cups of compost tea (see page 87) per foot of row. Spray with liquid seaweed in spring from bud set to just before full bloom.
Planting distance: 12–18 inches between plants and 3 feet between rows, or in single rows in raised beds.

AFTER PLANTING
Pick off all flower buds the first season on June-bearers. With day-neutrals, pick off flower buds in the first season till midsummer. (Alpines do not need to be flower-pruned.) Mulch all types with 6 inches of clean straw after planting. In fall, after the top inch of soil has frozen, cover the entire bed, including plants,

CONTAINER GROWING
For containers, grow strawberries that throw out few or no runners, such as the day-neutral or alpine strawberries. For a novelty, plant day-neutrals in hanging baskets — runners will fruit in midair.
   Use any potting mix suitable for houseplants. Discard old plants when their crowns turn woody and replace them with young plants. When winter comes, bring plants indoors if you want the harvest to continue. Give the plants abundant light and cool

temperatures. Otherwise, leave the containers outside, but protect them from cold.

From *Growing Fruits and Vegetables Organically* (Rodale Press, 1994)

with 4–6 inches of straw for winter protection. Remove mulch in spring when plants resume growth.

Watering: Apply 2 gallons per week per square foot of growing area. Water early in the day to facilitate rapid evaporation from leaves and fruit.

Pollinators: None required.

Harvesting

Time: June to early July for June-bearers; June through September for day-neutrals and alpines.

Test for ripeness: Pick when berries have turned bright red.

Years to bearing: 180 days to 1 year, depending on region and cultivar.

Yields per plant: June-bearers and day-neutrals — 1 pint; alpines — ¼ pint.

Propagation

Most cultivars produce runners, long stems with daughter plants developing where they touch the soil. The daughter plants can be transplanted to form new beds. Be extremely careful not to propagate from diseased plants.

▲ **'Red Gauntlet'** *is June-bearing.*

Pruning

Fruits develop on six-month-old plants to one- and two-year-old plants.

After planting: Pick off flowers. Prune off excess runners, as determined by your choice of growing system — matted row, spaced row or Hill system. See page 63.

Routine pruning: After harvesting June-bearing cultivars, renovate the bed by mowing plants to 2 inches above the ground. Rake off the leaves and mulch. Compost debris only if no viral diseases are present; otherwise destroy the debris.

Mature plants: Plants generally become diseased with age. Most gardeners avoid problems of disease buildup by tilling under beds that have been in place for three or four years and starting over with freshly purchased plants in a new bed.

Neglected plants: If young, mow and allow to regrow. Dig out oldest plants to allow room for new runners. Otherwise till under old plants.

PESTS, see page 83
Spider mites; strawberry root weevils (strawberry clippers); tarnished plant bugs.

DISEASES, see page 92
Powdery mildew; red stele; Verticillium wilt.

▼ **Grow 'Gento'** *for delicious-tasting fruit.*

RECOMMENDED CULTIVARS
Cultivars are highly region-specific. Check with your local Extension agent to learn which choices are best.

'Aptos' A day-neutral that grows well in California. High yields. Zones 6–9.

'Baron Solemacher' An old-fashioned alpine cultivar with high yields. Zones 4–8.

'Cardinal' Good June-bearing cultivar for the mid-Atlantic and southern states. Very disease resistant, with dark red berries suitable for fresh eating, freezing and preserves. Zones 6–9.

'Gento' Heavy yielding with tangy, good-flavored late-summer fruits. A good choice for limestone soils. Zones 5–8.

'Honeyoye' A high-yielding June-bearer for the Northeast, which tarnished plant bugs will pass over when other cultivars are available. Zones 4–8.

'Hood' A June-bearer for the Northwest with large berries. Mildew-resistant. Zones 5–9.

'Red Gauntlet' A good cropper and disease-resistant. June-bearing with large fruits. Zones 5–8.

'Sparkle' Fine flavor and good disease resistance. This is an excellent June-bearing cultivar for the Northeast and Midwest. Zones 4–8.

'Tristar' This day-neutral cultivar for the Northeast and Midwest is very disease-resistant and has sweet fruit. Zones 4–8.

Tayberries

One of the heaviest yielding of all brambles, tayberry's large purple fruits are very flavorful. The tayberry is a hybrid, with blackberry and raspberry parents. Its berries are hollow like those of a raspberry.

TYPE AND APPEARANCE

- Erect tayberry (*Rubus* hybrid)

Plants are deciduous, with perennial roots and upright biennial canes.

Height 6–8 feet; spread 3–5 feet; Zones 5–10.

Flowers: Pink, five petaled, 1 inch wide; in spring.

Leaves: Dark green, with three leaflets; dull red in fall.

PLANTING CONSIDERATIONS

When to plant: Spring or fall.

Where to plant: Full sun. Protect from winter winds.

Soil and feeding: Plants grow best in well-prepared soil enriched with organic matter. See page 28. Feed with 1– 1½ pounds of compost per foot of row in late winter each year.

Planting distance: 6 feet apart; 8 feet between rows.

AFTER PLANTING

Cultivate surrounding soil shallowly to eliminate weeds until midsummer. Then mulch with 6 inches of straw or weed-free hay.

Watering: Apply 2 gallons

per square foot of root spread every week until developed berries begin to color; then apply 1 gallon per square foot of root spread until frost. (The root area will be one and a half to two times wider than the top part of the plant.)

Pollinators: None required.

HARVESTING

Time: July and early August.

Test for ripeness: The dark purple-colored fruit should fall into your hands when gently pulled. Taste for ripeness.

Years to bearing: 1.

Yields per plant: 4–6 quarts.

PROPAGATION

Dig up suckers in early spring and transplant; see page 68. Only propagate disease-free plants.

PRUNING AND TRAINING

See raspberries, page 149.

PESTS, see page 84

Aphids; raspberry red-necked cane borers; raspberry fruitworms; raspberry sawfly; spider mites.

DISEASES, see page 93

Anthracnose; botrytis fruit rot; crown gall; crumbly berry virus; mosaic virus; powdery mildew; spur blight; Verticillium wilt.

RECOMMENDED CULTIVARS

Because tayberries are hybrids, cultivars and strains are not often identified. Buy from a nursery in your region to find a plant that will grow well in your backyard.

Attention required

Bee/butterfly friendly

Ornamental

Container growing

Yield

Ease of propagation

▼ **Sweet** *and juicy tayberries.*

USDA Plant Hardiness Zone Map

This map was revised in 1990 to reflect changes in climate since the original USDA map, done in 1965. It is now recognized as the best estimator of minimum temperatures available. Look at the map to find your area, then match its pattern to the key on the right. When you've found your pattern, the key will tell you what hardiness zone you live in. Remember that the map is a general guide; your particular conditions may vary.

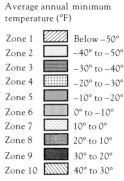

Average annual minimum temperature (°F)

Zone 1		Below −50°
Zone 2		−40° to −50°
Zone 3		−30° to −40°
Zone 4		−20° to −30°
Zone 5		−10° to −20°
Zone 6		0° to −10°
Zone 7		10° to 0°
Zone 8		20° to 10°
Zone 9		30° to 20°
Zone 10		40° to 30°

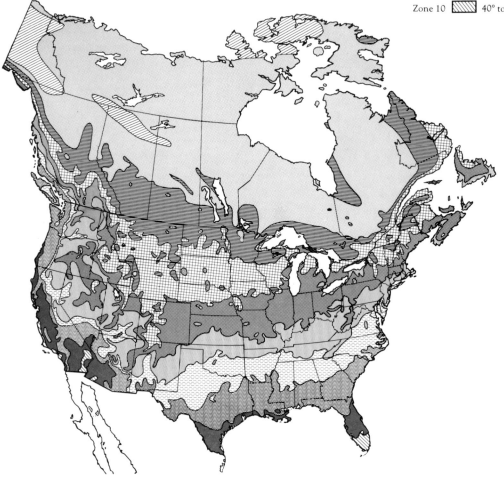

Resource Directory

SOURCES OF FRUITS AND BERRIES

Adams County Nursery, Inc.
P.O. Box 108
Aspers, PA 17304
Fruit trees

Ahrens Nursery & Plant Labs
P.O. Box 145
Huntingburg, IN 47542
Berry plants from tissue culture

Bear Creek Nursery
P.O. Box 411
Northport, WA 99157
Fruit trees, scionwood and budwood, budded trees, rootstocks

W. Atlee Burpee & Co.
300 Park Ave.
Warminster, PA 18974

C & O Nursery
P.O. Box 116
Wenatchee, WA 98801
Fruit trees

Champlain Isle Agro Associates
Isle LaMotte, VT 05463
Tissue culture bramble plants

Country Heritage Nursery
P.O. Box 536
Hartford, MI 49057
Mainly small fruits, some trees

Cumberland Valley Nurseries, Inc.
P.O. Box 471
McMinnville, TN 37110
Fruit trees

Edible Landscaping
P.O. Box 77
Afton, VA 22920
Specializes in fruits

Hastings
P.O. Box 115535
Atlanta, GA 30302
Specializes in plants for southern climates

Henry Leuthardt Nurseries, Inc.
P.O. Box 666
East Moriches, NY 11940
Specializes in small fruits

Hopkins Citrus and Rare Fruit Nursery
5200 S.W. 160th Ave.
Ft. Lauderdale, FL 33331

New York State Fruit Testing Cooperative Association Inc.
P.O. Box 462
Geneva, NY 14456–0462
Annual membership fee, new and old reliable fruits

North Star Gardens
19060 Manning Trail N.
Marine on St. Croix,
Specializes in berries

Raintree Nursery
391 Butts Rd.
Morton, WA 98356
Specializes in fruits and nuts

St. Lawrence Nurseries
R.R. 5, Box 324
Potsdam, NY 13676
Specializes in northern-hardy fruits and nuts

Southmeadow Fruit Gardens
Lakeside, MI 49116
More than 500 cultivars of two dozen fruits

Stark Brothers Nurseries & Orchards Co.
Hwy. 54
Louisiana, MO 63353
Fruits

Whitman Farms
3995 Gibson Rd. NW
Salem, OR 97304
Currants and gooseberries

GENERAL MATERIALS AND SUPPLIES FOR ORGANIC FRUIT GARDENERS

Harmony Farm Supply
P.O. Box 460
Graton, CA 95444

Necessary Trading Co.
One Nature's Way
Newcastle, VA 24127

INSECT TRAPS

Bio-Control Services
2600 Dalton St.
Ste. Foy
Quebec, Canada G1P 3S4

Consep, Inc.
213 S.W. Columbia
Bend, OR 97702–1013

Ladd Research Industries, Inc.
Box 1005
Burlington, VT 05402

Olson Products, Inc.
P.O. Box 1043
Medina, OH 44258

Orchard Equipment & Supply Co.
Box 540, Rte. 116
Conway, MA 01341

Trece, Inc.
P.O. Box 6278
1143 Madison Ln.
Salinas, CA 93907

BENEFICIAL INSECTS AND MITES

Beneficial Insectary
14751 Oak Run Rd.
Oak Run, CA 96069

Gerhert, Inc.
P.O. Box 39387
North Ridgeville, OH 44039

Integrated Orchard Management
617 S. Whitney St.
Visalla, CA 93277

Rincon-Vitova Insectaries, Inc.
P.O. Box 1555
Ventura, CA 93002

Stanley Gardens
P.O. Box 913
Belchertown, MA 01007

NEWSLETTERS

The Apple Press
Department of Plant and Soil
   Sciences
137 Hills Building
University of Vermont
Burlington, VT 05405

Fruit Notes
Department of Plant and Soil
   Sciences
205 Bowditch Hall
University of Massachusetts
Amherst, MA 01003

Northeast LISA Apple Newsletter
Department of Plant Pathology
Fernald Hall
University of Massachusetts
Amherst, MA 01003–2420

ASSOCIATIONS

North American Fruit Explorers
Rte. 1, Box 94
Chapin, IL 62628

BOOKS

Anderson, H.W. *Diseases of Fruit Crops.* New York: McGraw-Hill Book Co., 1956.

Bennett, Jennifer. *The Harrowsmith Book of Fruit Trees.* Willowdale, Ontario: Firefly Books, Ltd., 1991.

Bradley, Fern Marshall, and Barbara W. Ellis, eds. *Rodale's All-New Encyclopedia of Organic Gardening.* Emmaus, Pa.: Rodale Press, 1992.

Brickell, Christopher. *Pruning.* New York: Simon and Schuster, 1988.

Childers, Norman F. *Modern Fruit Science.* 9th, rev. ed. Gainesville, Fla.: Horticultural Publications, 1983.

Ellis, Barbara W., and Fern Marshall Bradley, eds. *The Organic Gardener's Handbook of Natural Insect and Disease Control.* Emmaus, Pa.: Rodale Press, 1992.

Galletta, Gene J. and David Himelrick. *Small Fruit Crop Management.* Englewood Cliffs, N.J.: Prentice Hall, 1990.

Hartmann, Hudson T., and Dala E. Kester. *Plant Propagation: Principles and Practices.* Englewood Cliffs, N.J.: Prentice Hall, 1983.

Hill, Lewis. *Pruning Simplified.* Updated ed. Pownal, Vt.: Storey Communications, 1986.

Hill, Lewis. *Secrets of Plant Propagation.* Pownal, Vt.: Storey Communications, 1985.

Horst, R. Kenneth. *Westcott's Plant Disease Handbook.* 5th ed. New York: Van Nostrand Reinhold Co., 1990.

Jones, A.L., and H.S. Aldwinckle. *Compendium of Apple and Pear Diseases.* St. Paul, Minn.: APS Press, 1990.

Joyce, David. *The Complete Guide to Pruning and Training Plants.* New York: Simon and Schuster, 1992.

Nick, Jean M.A., and Fern Marshall Bradley, eds. *Growing Fruits and Vegetables Organically.* Emmaus, Pa.: Rodale Press, 1994.

Ogawa, Joseph M. and Harley English. *Diseases of Temperate Zone Tree Fruits & Nut Trees.* Oakland, Calif.: ANR Publications, 1991.

Page, Stephen and Joseph Smillie. *The Orchard Almanac.* 2nd ed. Rockport, Maine: Spraysaver Publications, 1988.

Reich, Lee. *Uncommon Fruits Worthy of Attention.* Reading, Mass.: Addison-Wesley Publishing Co., Inc., 1991.

Scheer and Juergenson. *Approved Practices in Fruit and Vine Production.* Danville, Ill.: The Interstate Printers and Publishers, 1976.

Whealy, Kent, ed. *Fruit, Berry and Nut Inventory.* Decorah, Iowa: Seed Saver Publications, 1989.

Index

Credits

Key: *a* above, *b* below, *c* center, *l* left, *r* right

Heather Angel 116, 117; A–Z Botanical Collection 102, 123,
139; Rosalind Creasy 146; Derek Fell 2, 32, 41, 105, 106, 135, 141, 143;
Peter McHoy 75*l* & *r*, 101; Photo/Nats, Inc. 129;
Lynn Rogers/Hillstrom Stock Photo, Inc. 120; Harry Smith
Horticultural Collection 10*r*, 11, 12*a* & *b*, 13*a* & *b*, 24, 26, 27, 73,
74, 75*c*, 103, 107, 109, 110, 112, 114, 119, 121, 131, 132,
137, 145, 148, 149, 151, 152, 153; Unicorn Stock Photos 104,
(Betts Anderson) 18, (Kimberly Burnham) 23*ar*, (Wayne Floyd)
23*br*, (John L. Matthieson) 9, (Martha McBride) 14, 20, 67, 127,
134, (Karen Holsinger Mullen) 111, (Marshall Prescott) 23*al*,
(Charles E. Schmidt) 142, (Jim Shippee) 21, (Sue Vanderbilt)
23*br*, 125, 126; Wildlife Matters 30, 38, 147.

All other photography is copyright by Quarto.

While every effort has been made to acknowledge all copyright
holders, Quarto would like to apologize if any omissions have been
made.